An Introduction to Digital Signal Processing: A Focus on Implementation

An Introduction to Digital Signal Processing: A Focus on Implementation

Stanley H. Mneney

University of KwaZulu-Natal
Durban
South Africa

River Publishers

Aalborg

Published, sold and distributed by:
River Publishers ApS
PO box 1657
Algade 42
9000 Aalborg
Denmark
Tel.: +4536953197

ISBN: 978-87-92329-12-7
© 2008 River Publishers

Dedication

To my wife Edith, my daughter Thecla and my son Dan

Acknowledgments

The author would like to acknowledge the contribution made by Centre for TeleInFrastruktur at Aalborg University in Denmark in making their facilities available and for the support the center has provided in hosting the author during the development of this book. Special thanks to the Director of the centre Prof Ramjee Prasad who had the initial vision and encouraged and supported this work.

Stanley H Mneney
University of KwaZulu-Natal

About the Author

Prof S. H. Mneney
Pr. Eng., B.Sc.(Hons)Eng., M.A.Sc., Ph.D., SMSAIEE, MIET

Stanley Henry Mneney was born in Arusha in Tanzania and attended primary and secondary school in the same town. He completed the Cambridge O-level and the Tanzania National form 6 examinations with the maximum passes possible. He was admitted to the University of Science and Technology in Kumasi, Ghana, in 1972 to study Electrical Engineering and graduated in 1976, winning the Charles Deakens award for being the best engineering student in that year.

While under the employment of the University of Dar es Salaam, in Tanzania, he pursued a Master of Applied Science Degree at the University of Toronto in Canada and later did a sandwich PhD program between the University of Dar es Salaam and the Eindhoven University of Technology, in the Netherlands. In the early stages of his PhD work, he was awarded the 1984 Pierro Fanti International prize by INTELSAT and TELESPAZIO based on one of his publications.

Prof. S. H. Mneney has worked at the University of Dar es Salaam, the Universty of Nairobi, the University of Durban Westville, the University of Natal and is currently employed by the University of KwaZulu-Natal as Professor of Telecommunications and Signal Processing and is the current Head of School of Electrical, Electronic, and Computer Engineering. He is married with two children, now young adults.

He has been involved in the teaching of Electromagnetic Theory, Microwave Engineering, Digital Signal Processing and Telecommunications. His research interests include theory and performance of telecommunication systems, low cost rural telecommunications services and networks, Digital Signal Processing applications, and RF design applications using software and hardware.

Preface

In the past signal processing appeared in various concepts in more traditional courses like telecommunications, control, circuit theory, and in instrumentation. The signal processing done was analog and discrete components were used to achieve the various objectives. However, in the later part of the 20th century we saw the introduction of computers and their fast and tremendous growth. In the late 1960s and early 1970s a number of researchers resorted to modeling and simulation of various concepts in their research endeavors, using digital computers, in order to determine performance and optimize their design. It is these endeavors that led to the development of many digital signal processing algorithms that we know today. With the rapid growth of computing power in terms of speed and memory capacity a number of researchers wanted to obtain their results from near real-time to real time. This saw the development of processors and I/O devices that were dedicated to real-time data processing though initially at lower speeds they are currently capable of processing high speed data including video signals. The many algorithms that were developed in the research activities combined with software and hardware that was developed for processing by industry ushered in a new course into the Universities curricula; Digital Signal Processing.

For many years the course Digital Signal processing was offered as a postgraduate course with students required to have a background in telecommunications (spectral analysis), circuit theory and of course Mathematics. The course provided the foundation to do more advanced research in the field. Though this was very useful it did not provide all the necessary background that many industries required; to write efficient programs and to develop applications. In many institutions a simplified version of the postgraduate course has filtered into the undergraduate programme. In many cases that we have examined this course is a simplified version of the postgraduate course, it is very theoretical and does not pass the necessary tools to students that industry requires.

This book is an attempt to bridge the gap. It is aimed at undergraduate students who have basic knowledge in C programming, Circuit Theory, Systems and Simulations, and Spectral Analysis. It is focused on basic concepts of digital signal processing, MATLAB simulation and implementation on selected DSP hardware. The candidate is introduced to the basic concepts first before embarking to the practical part which comes in the later chapters.

Chapter 1 introduces the students to discrete-time signals and systems hopefully for the first-time. It shows how such signals are represented and related through the sampling process. Some applications are introduced and the motivation for digital signal processing is given. Chapters 2, 3, and 4 introduce the concept of the transform domain. The reason for sampling a continuous spectrum of discrete-time signals is developed and the speeding up of computations using the Fast Fourier Transform is elucidated. Chapter 4 introduces an important tool the z-transform that is used to present, analyze, and manipulate DSP structures.

It is important that the students are able to design analog filters as this is the starting point for some type of digital filters. Chapter 5 provides the necessary background to achieve this goal. The use of MATLAB in the design is also introduced. Chapter 6 deals with the design of digital filters. There are many different design methods but in this book we focus on only the most common methods.

Chapter 7 deals with implementations issues in the processors. It deals with problems related to quantization of signal variables and coefficients, number representations and problems of overflow. Chapter 8 deals with existing implementation hardware and focuses on the most common hardware used in academic institutions and industry. The student is introduced to Code Composer Studio. We have noticed that most of the students assigned to do DSP projects are not aware of the existence of these features and therefore do not make use of them. Chapter 9 winds up the book with a number of implementation examples.

Each chapter ends with some theoretical and in some cases practical problems. At the end of Chapter 9 are some proposed projects. The book is recommended for use at the final year of the undergraduate electrical engineering programme for a one semester course. It is essential that the necessary equipment is made available.

Contents

1

Introduction to Digital Signal Processing

1.1 A Brief Introduction to Digital Signal Processing

Signal processing is simply the manipulation of the properties of a specific signal to obtain a signal with more desirable properties. Properties such as amplitude, phase, or frequency spectrum may be altered to meet a specific requirement. In the early days electronic engineers achieved signal processing using discrete hardware components such as resistors, capacitors, inductors, transistors, diodes, and other semiconductor devices. In such a case a signal variable that was continuous with time was used as an input to a hardware device that produced a new version of the signal variable where some of the properties have been altered. In digital signal processing the processes that were achieved using hardware are done using software.

In order to process signal variables that are continuous with time using software the variables have to be converted to the correct format; normally a sequence of numbers. This is done using analog to digital converters. A processor would then manipulate the signal in some desired fashion. After going through the processors the resulting sequence of numbers has to be converted back to analog using digital to analog converters. In the dawn of digital signal processing the processors were slow and the applications of digital signal processing were limited. Today, we have very fast and power efficient processors that the applications of digital signal processing have increased dramatically.

In order to understand digital signal processing it is important to first look at signal classification in the time-domain.

1.2   Signal Classification

Signals can be classified in terms of the continuity of the independent and dependent variables as follows:

  (i) An analog signal: The independent and the dependent variables defining the signal are continuous in time and amplitude. This means that for each specified time instance, the signal has a specified amplitude value.
 (ii) Continuous-time signal: The time variable is continuous in the range in which the signal is defined. If the signal variable is represented by x, time variable is t such a signal is denoted as $x(t)$.
(iii) Discrete-time signal: The time variable is discrete in the range in which the signal is defined. If the signal variable is x and the time variable has been sampled at time instances n, where $n = n'T$ then the signal is denoted as $x(n)$. A discrete-time signal is also referred to as a sampled signal since it is obtained by directly sampling a targeted signal. It should be noted that the amplitude of the sampled signal can take any value within a specified amplitude range and we therefore say that the amplitude of discrete-time signal is continuous.
(iv) A digital signal: This is a signal that is discrete in time and discrete in amplitude. It is represented in the same way as a discrete-time signal.

Signals can also be classified in terms of the predictability of the dependent variables with respect to the independent variable as follows:

  (i) A signal is said to be deterministic if the dependent variable is predictable at any instance of the independent variable time. A deterministic signal can be expressed by an explicit mathematical expression.
 (ii) A random signal, on the hand, has an unpredictable dependent variable at any instance of the independent variable time. Such a signal can only be defined in terms of its statistical properties.

All the above classifications of digital signals can further be classified in terms of their dimensionality. Here, we will only elaborate this classification using discrete-time sequences and we will leave the rest to the student.

(i) A one-dimensional signal has only one-independent variable and one-dependent variable. A discrete-time signal $x(n)$ is a one-dimensional signal as it has only one-independent variable, discrete-time (n), and one-dependent variable, the amplitude of $x(n)$.

(ii) A two-dimensional signal has two-independent variables and one-dependent variable. The samples n and m are taken in the spatial domain. The two-dimensional signal is discrete in the spatial domain in two-dimensions. The independent variables are n, m which define the dependent variable $x(n,m)$. A good example is a photographic image where n, m define the spatial location and $x(n,m)$ defines the grey level at the location.

(iii) A three-dimensional signal has three-independent variables and one-dependent variable. A discrete-time signal $x(n,m,\tau)$ is a three-dimensional signal as it has two-independent variable in the spatial domain (n,m) and one-independent variable τ in the time domain. The three-independent variables define the one-dependent variable, the intensity of $x(n,m,\tau)$. An example of a three-dimensional signal is video signal where a signal at spatial location (n,m) is changing with respect to time τ.

A system can be classified as analog, discrete-time or digital systems depending on the type of signals they handle. An analog system would produce an analog signal from an analog input signal. On the other hand, a discrete-time or digital system can produce a discrete-time or digital signal from a discrete-time or digital signal. However, a discrete-time or digital system can produce an analog signal from an analog input with the aid of ADC and DAC. From now on we will not distinguish between discrete-time signals and digital signals as these are handled in the same way. We will also not distinguish between digital systems and discrete-time systems as there is technically no difference between them.

Many natural phenomena produce analog signals and signal processors are inherently digital systems. Thus in order to process analog signals with digital processors the analog signals must be converted to digital. The process whereby analog signals are converted to digital signals involves sampling and quantization. In the next section, we will discuss the sampling process.

1.3   The Sampling Process

Sampling a continuous-time signal implies taking snap shots of the signal at specific instances of time as shown in Figure 1.1.

It is easy to note that if we take very few samples we will not be able to obtain the original waveshape by interpolation as shown in Figure 1.2.

If we sample at a rate similar or higher than that shown in Figure 1.1 it is possible to reproduce a wave shape almost identical to the original wave shape. If we sample at higher rates we generate more samples and hence we create a much larger demand for memory to store the samples. We can represent the sampling process mathematically. Sampling is a process where an analog signal is multiplied by an impulse train. Figure 1.3(a) represents an analog input signal $x(t)$ that is to be sampled and Figure 1.3(b) represents an impulse train which

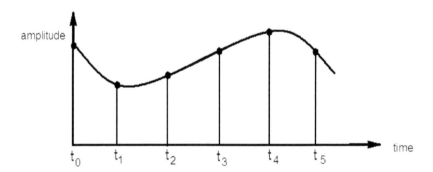

Fig. 1.1 The sampling process.

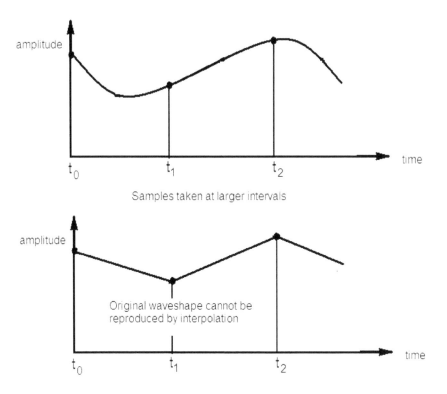

Samples taken at larger intervals

Original waveshape cannot be
reproduced by interpolation

Fig. 1.2 Impact of sampling at large time intervals.

mathematically is represented by Eq. (1.1).

$$s(t) = \sum_{n=-\infty}^{\infty} \delta(t - nT_s).$$ (1.1)

The sampled signal is given by $y(t)$ and is represented by Eq. (1.2) as follows:

$$y(t) = x(t) * s(t) = x(t) * \sum_{n=-\infty}^{\infty} \delta(t - nT_s) = \sum_{n=-\infty}^{\infty} x(nT_s)\delta(t - nT_s).$$
(1.2)

In the sampled signal therefore the location of the samples are determined by the impulse train and its weight is determined by the value of the analog signal at the specific instance.

We will get a clearer picture if we look at sampling in the frequency domain. Suppose the spectrum of $x(t)$ is given by $X(f)$ as shown in

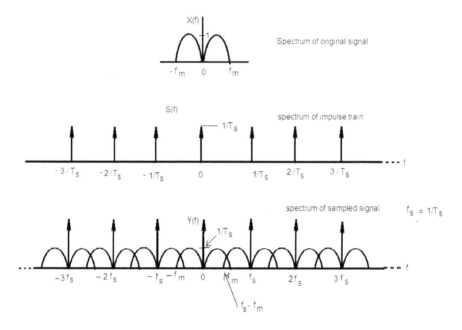

Fig. 1.5 The sampling process in the frequency domain for $f_s < 2f_m$.

are a number of claims from independent researchers including E. T. Whittaker, H. Nyquist, K. Kupfmuller, and V. A. Kotelnikov. Claude Shannon proved the theorem in 1949. The theorem has been referred to in literature under various names and the most common are the Nyquist theorem and the Shannon's theorem. We will here refer to it simply as the sampling theorem and it stated as follows:

The Sampling Theorem.   For any baseband signal that is bandlimited to a frequency f_m, the sampling rate f_s must be selected to be greater or equal to twice the highest frequency f_m in order for the original baseband signal to be recovered without distortion using an ideal lowpass filter with a cut-off frequency f_c such that $f_m \leq f_c \leq f_s - f_m$.

In Figure 1.5, we notice that the "tails" and the "heads" of the spectral components of adjacent periods overlap. In an attempt to recover the baseband signal using an ideal lowpass filter it is impossible to remove the error introduced by the spectral components from the adjacent period. The inherent error is referred to as the *aliasing error*. In order to eliminate the aliasing error the baseband signal must be properly bandlimited and sampling done according to the sampling theorem. The bandlimiting filters that are used are referred to as *anti-aliasing filters.*

In practice when a sample is taken the sample value is maintained constant until when the next sample is taken. This ensures that the input value to the processor does not change during the sample period as the reading of the input can be done at any time during this period. A "sample and hold" circuit is normally used for this function. The input circuit of a DSP processor has essentially the elements shown in Figure 1.6; an anti-aliasing filter, a sample and hold circuit and ADC circuit. The ADC circuit quantizes the output of the sample and hold. In the simple case of Figure 1.6 the ADC can accept only four quantization levels. For this quantizer any voltage level that is between two quantization intervals is truncated to the lower level. This is referred to as quantization by *truncation*. It is possible for quantization to be done by rounding to the nearest level. Each level is coded using 2 bits in a four level quantization system. The output of the ADC is therefore binary as shown in Figure 1.6.

Many ADCs available in the market also contain the sample and hold circuit in the same integrated circuit. The four most popular ADCs technologoies are the successive approximation ADC, the Dual slope ADC, the flash ADC, and the Sigma delta ADC. The flash ADC converts the input signal faster than other ADCs. The successive approximation, dual slope and the flash ADCs are based on the principles of sampling around the Nyquist frequency and thus require good anti-aliasing filters. They achieve high resolution using precise component matching and laser trimming. The sigma delta ADC, on the other hand

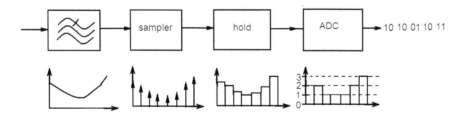

Fig. 1.6 The input stage of a digital signal processor.

use a low resolution ADC (one bit quantizer) with a sampling rate many times higher than the Nyquist rate.

The output of the processor is binary and may be required to be converted to an analog signal. This is achieved using a DAC device. The DAC devices are simpler and lower in cost compared to the ADC devices. The most common form of the DAC devices are the multiplying devices; both the current source multiplying DAC and voltage source multiplying DAC. The multiplying DAC devices are fast with settling times of about 100 ns or less. Most of the commercially available DAC are "Zero-Order-Hold"; they convert the binary input to an analog level and hold until the next sample arrives. This results "staircase" waveform and requires the use of a reconstruction filter to smoothen the waveform, Figure 1.7.

We have looked at the peripheral input and output circuits required in the implementation of a DSP system. These two circuits feed into and out of the DSP processor. The processor itself may contain several units such as the arithmetic logic unit (ALU), volatile memory unit for data and program (RAM), non-volatile memory units (ROM), and buses

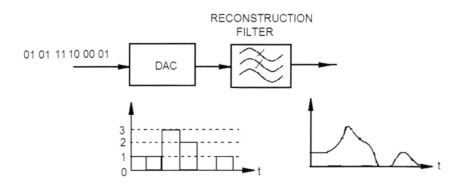

Fig. 1.7 The output stage of a Digital Signal Processor.

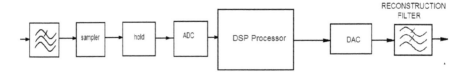

Fig. 1.8 The essential elements of a DSP system.

or conduits of data and programs. The essential elements of a DSP system are thus shown in Figure 1.8, but these may vary depending on the particular application.

1.4 Discrete-Time Signals

Discrete-time signals are a set of numbers or sequences formed from a set of samples taken from an analog signal or they may be words taken from digital binary data. Each word or sample may be represented by a variable $x(n)$, where n is a time index and is always an integer which is defined in a specified range. The sequence or the discrete-time signal itself is denoted by $\{x(n)\}$, with parenthesis used to differentiate the sequence from the individual sample $x(n)$. Where there is no ambiguity between the sample and the sequence the parenthesis is dropped.

1.4.1 Examples of Discrete-Time Signals

(i) The unit sample sequence (see Figure 1.9)

$$\delta(n) = \begin{cases} 1 & \text{for } n = 0 \\ 0 & \text{otherwise.} \end{cases} \tag{1.5}$$

(ii) The unit step sequence (see Figure 1.10)

$$\mu(n) = \begin{cases} 1 & \text{for } n \geq 1 \\ 0 & \text{otherwise.} \end{cases} \tag{1.6}$$

(iii) The discrete-time sinusoidal sequence

$$x(n) = \sin(\omega_0 n) \quad \text{for } -\infty < n < \infty, \tag{1.7}$$

where ω_0 is a normalized digital frequency in radians per sample (see Figure 1.11).

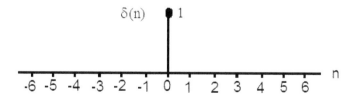

Fig. 1.9 The unit sample sequence.

Fig. 1.10 The unit step sequence.

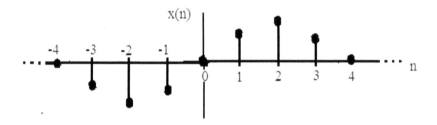

Fig. 1.11 Discrete time sinusoidal sequence.

(iv) The discrete time exponential sequence

$$x(n) = A\alpha^n \mu(n). \tag{1.8}$$

When multiplied by a step sequence we make the exponential sequence zero for $n < 0$ and $A\alpha^n$ for $n \geq 0$. The plot of the exponential sequence for $A = 32$ and $\alpha = 0.5$ is given in Figure 1.12.

The sequences introduced above (except the unit sample sequence) are either single sided or double sided infinite length sequences. It is possible also to have sequences that are of finite length. A typical example is given below.

(v) This example shows a finite length sequence of length 8. Here $x(0) = 2.0$.

$$x(n) = \{1.0, 2.2, 1.5, 2.0, -3, -1.5, -0.5, 1.2\}$$
$$\text{for } -3 \leq n \leq 4. \tag{1.9}$$

1.4.2 Arithmetic Operation on Sequences

There are a number of arithmetic operations that can be applied to sequences. In the paragraph below we show the arithmetic operations

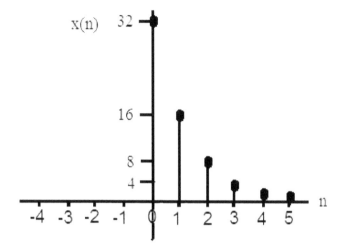

Fig. 1.12 Discrete-time exponential sequence.

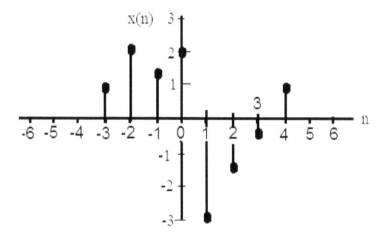

Fig. 1.13 A finite length sequence.

and how they are represented in a signal flow graph:

(i) *Multiplication*: The multiplication of two sequences $x(n)$ and $y(n)$ to form a new sequence $z(n)$ is achieved by multiplying corresponding samples. If the sequences are not of the same length they are padded with zeros to make them defined in

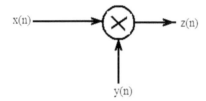

Fig. 1.14 Multiplication of two sequences.

exactly the same range

$$z(n) = x(n) \cdot y(n). \tag{1.10}$$

Schematic representation is shown in Figure 1.14

(ii) *Multiplication by a constant*: The multiplication of a sequence $x(n)$ by a constant α to obtain a new sequence $y(n)$ is achieved if each sample of the sequence $x(n)$ is multiplied by the same constant

$$y(n) = \alpha x(n) \quad \text{for all } n. \tag{1.11}$$

Schematic representation is shown in Figure 1.15

(iii) *Addition*: The addition of two sequences $x(n)$ and $y(n)$ to form a new sequence $z(n)$ is achieved by adding corresponding samples. If the sequences are not of the same length they are padded with zeros to make them defined in exactly the same range

$$z(n) = x(n) + y(n) \quad \text{for all } n. \tag{1.12}$$

Schematic representation is shown in Figure 1.16

(iv) *Delay operation*: A sequence $x(n)$ is delayed by one unit delay to give a new sequence $y(n)$ if each sample of $x(n)$ is delayed

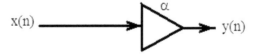

Fig. 1.15 Multiplication by a constant.

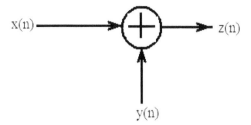

Fig. 1.16 Addition of two sequences.

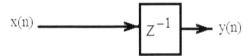

Fig. 1.17 Delay operation.

by the same unit delay. A unit delay is represented by a multiplication by Z^{-1} in the z-domain.

$$y(n) = x(n-1). \tag{1.13}$$

Schematic representation is shown in Figure 1.17

(v) *Pick off node*: This is a node where copies of the original sequence are generated. Schematic representation is shown in Figure 1.18

A combination of these operations can be obtained from the following example. This example shows that the current output is given by a linear combination of the current input and past inputs. It is shown schematically in Figure 1.19.

$$y(n) = \alpha_0 x(n) + \alpha_1 x(n-1) + \alpha_2 x(n-2)$$
$$+ \alpha_3 x(n-3) + \alpha_4 x(n-4). \tag{1.14}$$

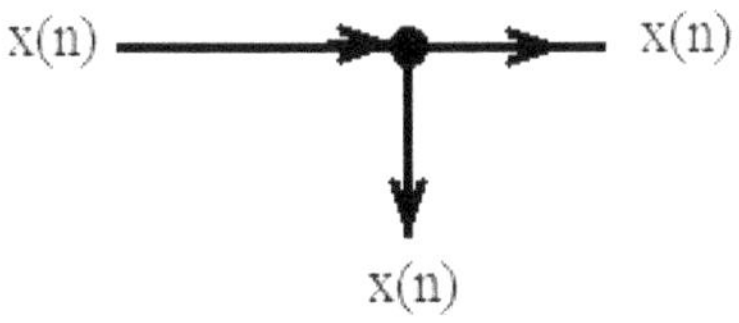

Fig. 1.18 Copying operation.

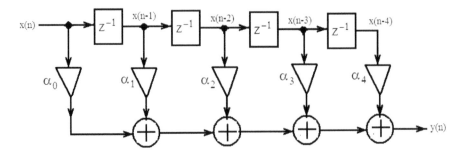

Fig. 1.19 A combination of basic operations.

1.5   Discrete-Time Systems

The combination of basic operations that formed Figure 1.19 represents a digital system. Depending on how the coefficients are selected the system can be a digital filter or an equalizer. This system can be implemented in digital hardware or in software. In the sections that follow we will look at a few examples of simple systems.

(i) The Moving Average Filter is represented by the following equation $y(n) = \frac{1}{M} \sum_{k=0}^{M-1} x(n-k)$, where $x(n)$ is the input and $y(n)$ is the out sequence.

   Here we show the impact of the moving average filter. Noise $d(n)$ generated from random noise generator function is added on to the signal $s(n) = 2\sin(wn)$ to give

$$x(n) = s(n) + d(n). \tag{1.15}$$

This is shown in Figure 1.20. The resulting signal $x(n)$ is then applied to a moving average filter of length 4 and the result is shown in Figure 1.21.

(ii) A Differentiator: Figure 1.22 shows a derivative $y(t)$ given by the slope of $x(t)$ at time t_0. The slope can be approximated by $y(nT) = \frac{x(nT) - x((n-1)T)}{(nT) - (n-1)T}$. If we define $T = 1$ then we can approximate differentiation using a difference equation as

$$y(n) = \frac{1}{T}\{x(n) - x(n-1)\}. \tag{1.16}$$

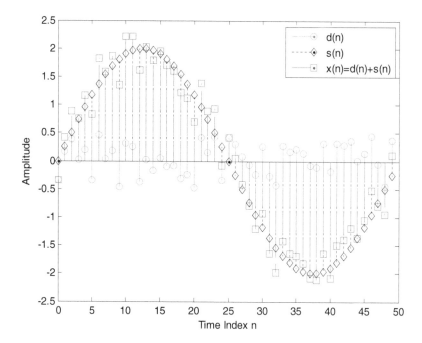

Fig. 1.20 Random noise added to sinusoidal signal.

The differentiator expressed in a form of a difference equation is an example of a discrete-time system (see Figure 1.23).

(iii) An Integrator: Integration can be represented by Eq. (1.17) as shown in the graph of Figure 1.24.

$$y(t) = \int_{t_1}^{t_2} x(t)dt. \tag{1.17}$$

Using the trapezoidal rule integration can be approximated by

$$y(nT) = T \left\{ \frac{x(nT) + x((n-1)T)}{2} \right\}.$$

If we let $n = n'T$ we can approximate integration using a difference equation as follows:

$$y(n) = \frac{T}{2} \{x(n) + x((n-1)T)\} + y(n-1). \tag{1.18}$$

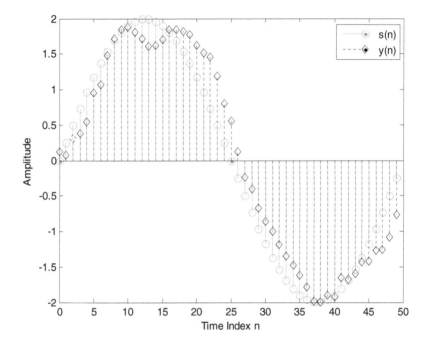

Fig. 1.21 The impact of the moving average filter on the noisy signal.

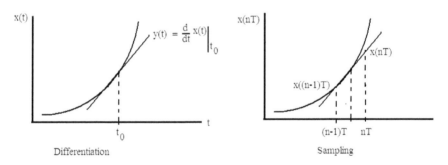

Fig. 1.22 Differentiation operation and approximation.

The approximation to the integrator using a linear difference equation is an example of a simple discrete-time system. The term $y(n-1)$ gives the initial condition (see Figure 1.25).

(iv) An up-sampler: The up-sampler is a discrete-time system that will increase the sampling rate of a sequence. It does this by inserting zero value samples in between samples of an

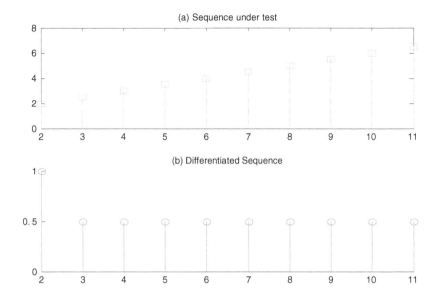

Fig. 1.23 Input and Output of a discrete-time differentiator.

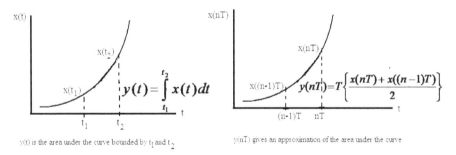

Fig. 1.24 Integration operation and approximation.

existing sequence (see Figure 1.25). The up-sampled sequence $y(n)$ of an input sequence $x(n)$ is defined as

$$y(n) = \begin{cases} x(n/L), & n = 0, \pm L, \pm 2L, \ldots \\ 0 & \text{otherwise.} \end{cases} \qquad (1.19)$$

(v) A down sampler: The down-sampler is a discrete-time system that will decrease the sampling rate of a sequence (see Figure 1.27). It does this by selecting only the Mth

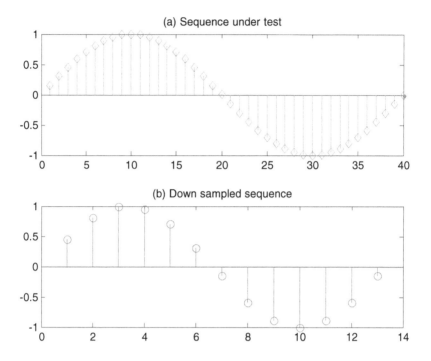

Fig. 1.27 Down sampling by a factor of 3.

independent of the time at which the input is applied. Mathematically we can state that the system is time invariant if the response to the input $x(n)$ is $y(n)$ then the response to the input $x(n - n_0)$ is $y(n - n_0)$.

A system that satisfies both the linearity property and the time-invariance property is referred to as a Linear-Time-Invariant (LTI) system. LTI systems are easy to model and analyze. The input output relationship can be represented by a convolution process.

(iii) *Causality Property*

A system is said to be causal if its output does not depend on future inputs. Mathematically we can say that at the time n_0 the output $y(n)$ depends on $x(n)$ for $n \leq n_0$ only.

This property can further be extended to cover the general situation for causal systems that whatever happens at the

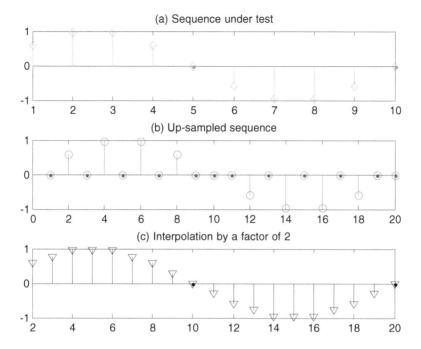

Fig. 1.28 Interpolation by a factor of 2.

output must be preceded by what happens at the input. That is changes in the input must precede changes in the output. There can be no output before an input is applied.

(iv) *Stability Property*

A system is said to be stable if and only if a bounded input produces a bounded output. This type of stability is referred to as Bounded Input Bounded Output (BIBO) stability.

1.7 Some Applications of Digital Signal Processing

In the previous sections, we have introduced the discrete-time signals and systems. We have seen how to obtain the discrete-time signal from an analog signals through the sampling process. In this section, we will briefly list the reasons for preferring digital signal processing over analog signal processing before we look at some applications.

The key reasons for use of digital signal processor are

(i) Applications are normally in the form of programs. One can have the same hardware to process different applications. In order to process a different application one simply loads the relevant program. In analog signal processing different applications can only be processed with dedicated software.

(ii) If the same application is run in two different signal processors the results are identical. This is not the case when using analog systems. The components making analog system are not identical due to tolerance, aging, and heating. These variables do not affect software programs.

(iii) Digital systems are more stable than analog systems. Again the discrete components that make analog systems change in value due to heating, aging or influence of humidity. This change causes a deterioration of the performance of the system. This rarely happens to digital systems because the greater part of a digital system is simply software.

(iv) It is possible to implement special applications with digital systems that is not possible with analog systems. Examples of these include the implementation of linear phase filters, adaptive filters, and notch filters. To implement these with analog systems will be very difficult or almost impossible.

The key reasons for still using analog signal processors are

(i) There is additional complexity in digital systems due to the need for ADC and DAC converters. When two Digital Signal Processors communicate there is the added problem of synchronization.

(ii) The additional circuits in (i) have an extra power requirement. It is possible to build analog filters using components that do not require any power supply.

(iii) Whereas analog systems can be built to operate at very high frequencies the best DSP processors can handle between 1 and 2 GHz.

Despite the demerits of DSP processors listed above we have seen almost an exponential growth in their application. Below is a short list only of DSP applications

(1) In high fidelity music reproduction. A good example is reproduction of sound from a compact disc where the following can be achieved digitally: Eight-to-four decoding (EFM), Reed Solomon coding, and rate conversion. The rest of the processes are analog; DAC conversion, Bessel filtering, and power amplification. DSPs are used in the reproduction of special sound effects in music systems.

(2) Reproduction of synthetic speech where recorded speech can be altered to sound differently for games or camouflage.

(3) Image processing where exotic images can be converted to more conventional images that can be easily evaluated by the eye. There are image processing procedures that result in image enhancement (like sharpening the edges), reducing random noise, correcting image blur and motion distortion.

(4) DSPs are used in all cellular phones for link control, power control, channel selection, roaming, sim-card operations; such as authentication and registration. DSPs are used in nearly all advanced instrumentation; seismic detectors, modern radar applications, and in medical equipment.

1.8 Problems

1.7.1 An analog signal made from three sinusoidal components is given by $x(t) = 3\sin(7\Omega_0 t) + 5\sin(5\Omega_0 t) + 7\sin(11\Omega_0 t)$. The signal is sampled to form a discrete-time signal.

(i) What is the recommended sampling frequency?

(ii) Give an expression for the discrete-time signal formed.

(iii) Explain the consequence of sampling at a frequency of $10\Omega_0$.

1.7.2 A discrete-time signal formed by a combination of three sinusoidal components is given by $x(n) = 2\sin(w_0 n) + 3\sin(3w_0 n) + 4\cos(4w_0 n)$.
Determine

   (i) the period of each term on the left-hand side, and

   (ii) the period for the three components together.

1.7.3 Express the following discrete-time sequences as a linear combination of the unit sample sequence and their delayed versions

   (i) $x_1(n) = [1, 2, -3, 0, 4, 0, 0, 2]$ for $-3 \le n \le 4$, and

   (ii) $x_2(n) = [1, 2, -3, 0, 4, 0, 0, 2]$ for $3 \le n \le 10$

1.7.4 Express the following discrete-time sequences as a linear combination of the unit step sequence and their delayed versions

   (i) $x_1(n) = [1, 2, -3, 0, 4, 0, 0, 2]$ for $-3 \le n \le 4$, and

   (ii) $x_2(n) = [1, 2, -3, 0, 4, 0, 0, 2]$ for $3 \le n \le 10$.

1.7.5 An exponential sequence is given by $x(n) = A\alpha^n \mu(n)$ for $\alpha < 1$.
Determine

   (i) whether $x(n)$ is absolutely summable, and

   (ii) the energy in the sequence $x(n)$.

1.7.6 Develop an expression between the input and output of the following discrete-time systems.

(i)

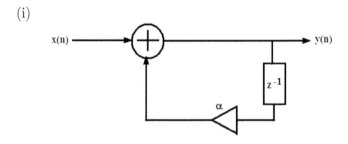

(ii)

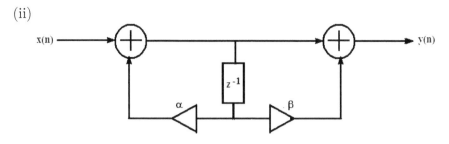

(iii)

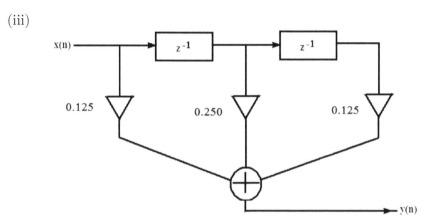

1.7.7 Show whether a moving average system is
    (i) linear (ii) stable (iii) causal, and (iv) time invariant.

1.7.8 Show whether a discrete time differentiator is
    (i) linear (ii) stable (iii) causal, and (iv) time invariant.

1.7.9 An output $y(n)$ from a discrete-time sequence $x(n)$ is
    given by a convolution process represented by $y(n) = \sum_{k=-\infty}^{\infty} h(k)x(n-k)$, where $h(n)$ is the impulse response to
    the system. Show that the condition for stability of the sys-
    tem is that impulse response has to be absolutely summable.

1.7.10 Show how the integral $y(t) = \int_0^t x(\tau)d\tau$ can be approximated
    by linear constant coefficients difference equations.

1.7.11 Compute the energy of the following sequence $x(n) = \sin(\omega_0 n)$ for $0 \leq n \leq N-1$, where N is the length of the
    sequence.

2

The Transform Domain Analysis: The Discrete-Time Fourier Transform

2.1 The Discrete-Time Fourier Transform

The Fourier Transform of a continuous-time signal is given by

$$X(\Omega) = \int_{-\infty}^{\infty} x(t)e^{-j\Omega t}dt, \tag{2.1}$$

where Ω represents the analog frequency in radians/s. If $x(t)$ is sampled at a sampling frequency $1/T$ then the integration becomes a summation over the variable n as

$$X(e^{j\Omega T}) = \sum_{n=-\infty}^{\infty} x(nT)e^{-j\Omega nT}. \tag{2.2}$$

We can express ΩT as $\Omega T = 2\pi f/f_s = \omega$, where $f_s = 1/T$ and ω is defined as the normalized digital frequency in units of radians per sample. If we define $x(n)$ to represent sample $x(nT)$ we obtain the expression

$$X(e^{j\omega}) = \sum_{n=-\infty}^{\infty} x(n)e^{-j\omega n}. \tag{2.3}$$

Equation (2.3) is the known as the Discrete-time Fourier Transform (DTFT) of the sequence $x(n)$. Note that the DTFT is a complex function and can be expressed as a magnitude and phase as in Equation (2.4).

$$X(e^{j\omega}) = |X(e^{j\omega})|e^{j\varphi}. \tag{2.4}$$

If we expand Equation (2.3) we obtain the real part and the imaginary part of the DTFT

$$X(e^{j\omega}) = \sum_{n=-\infty}^{\infty} x(n)\cos(\omega n) + j \sum_{n=-\infty}^{\infty} x(n)\sin(\omega n)$$

$$= \mathrm{Re}(X(e^{j\omega}) + \mathrm{Im}(X(e^{j\omega}). \tag{2.5}$$

The phase angle of the DTFT is then given by

$$\varphi = \tan^{-1}\left(\frac{\mathrm{Im}(X(e^{j\omega}))}{\mathrm{Re}(X(e^{j\omega})}\right). \tag{2.6}$$

Since $\sin(\omega n)$ is an odd function with respect to frequency the imaginary part of the DTFT is always an odd function with respect to frequency. Also since $\cos(\omega n)$ is an even function with respect to frequency the real part of the DTFT is always even with respect to frequency. The quotient of an odd function with an even function is an odd function and therefore the phase angle φ is always an odd function.

The magnitude of the DTFT can be obtained from the expression in Equation (2.7)

$$|X(e^{j\omega})|^2 = (\mathrm{Re}(X(e^{j\omega}))^2) + (\mathrm{Im}(X(e^{j\omega}))^2). \tag{2.7}$$

From Equation (2.7), $|X(e^{j\omega})| = |X(e^{-j\omega})|$ making the magnitude an even function.

From the above analysis we can conclude that the magnitude of the DTFT is an even function, the phase angle of the DTFT is an odd function, the real part of the DTFT is an even function and the imaginary part of the DTFT is an odd function.

The phase is defined for the interval $-\pi \le \varphi < \pi$ because if we shift the phase by $2\pi k$ we get $e^{j\varphi + 2\pi k} = e^{j\varphi}$. This implies that the phase function is ambiguous, it cannot be specified uniquely for each DTFT.

The existence of the DTFT depends on whether $X(e^{j\omega})$, as given by Equation (2.3), converges or not. We can express a sufficient and necessary condition for the existence of the DTFT as

$$\lim_{k \to \infty} |X(e^{j\omega}) - X_k(e^{j\omega})| = 0, \tag{2.8}$$

where $X_k(e^{j\omega}) = \sum_{n=-k}^{k} x(n)e^{-j\omega n}$.

A sufficient condition for the existence of a DTFT is that the absolute value of the DTFT has to be bounded. That is

$$|X(e^{j\omega})| = \left| \sum_{n=-\infty}^{\infty} x(n)e^{-j\omega n} \right| \leq \sum_{n=-\infty}^{\infty} |x(n)| < \infty. \qquad (2.9)$$

This implies that

$$\sum_{n=-\infty}^{\infty} |x(n)| < \infty. \qquad (2.10)$$

A sufficient condition for the existence of the DTFT is the sequence $x(n)$ must be absolutely summable.

Here we will show some DTFTs of selected sequences.

Example 2.1. Determine the DTFT of the unit sample sequence
$x(n) = \delta(n) = \begin{cases} 1 & \text{for } n = 0 \\ 0 & \text{elsewhere.} \end{cases}$

Solution

$$X(e^{j\omega}) = \sum_{n=-\infty}^{\infty} x(n)e^{-j\omega n} = \sum_{n=-\infty}^{\infty} \delta(n)e^{-j\omega n} = 1.$$

Here the DTFT or the frequency spectrum of the unit sample sequence is real and constant and with zero phase in the interval $-\pi \leq \varphi < \pi$.

Example 2.2. Determine the DTFT of a causal exponential sequence $x(n) = \alpha^n \mu(n)$ for $\alpha < 1$.

Solution

$$X(e^{j\omega}) = \sum_{n=-\infty}^{\infty} x(n)e^{-j\omega n} = \sum_{n=-\infty}^{\infty} \alpha^n \mu(n)e^{-j\omega n}$$

$$= \sum_{n=0}^{\infty} \alpha^n e^{-j\omega n} = \sum_{n=0}^{\infty} (\alpha e^{-j\omega})^n$$

$$= \frac{1}{1 - \alpha e^{-j\omega}} \quad \text{since } |\alpha e^{-j\omega}| < 1.$$

In order to extract the magnitude and phase angle we proceed as follows:

$$X(e^{j\omega}) = \frac{1}{1 - \alpha e^{-j\omega}} = \frac{1}{1 - \alpha\cos\omega + j\alpha\sin\omega}$$

$$|X(e^{j\omega})|^2 = \frac{1}{1 - 2\alpha\cos\omega + \alpha^2}$$

and the phase angle

$$\varphi = \tan^{-1}\left(\frac{\alpha\sin\omega}{1 - \alpha\cos\omega}\right).$$

The plots of the magnitude and phase response are shown in Figure 2.1 for $\alpha = 0.5$. The program for plotting is given in the appendix as Program 2.2.

Example 2.3. Determine the DTFT of an anti-causal exponential sequence $x(n) = -\alpha^n\mu(-n-1)$ for $\alpha > 1$.

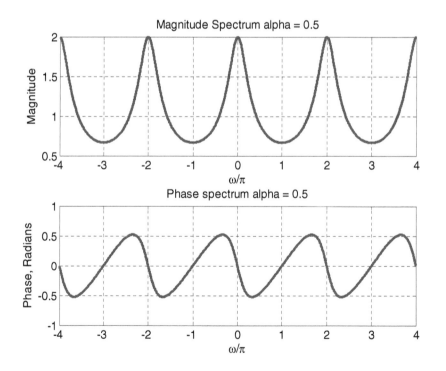

Fig. 2.1 Plot for the magnitude and phase response for $x(n) = \alpha^n\mu(n)$ for $\alpha < 1$.

Solution

$$X(e^{j\omega}) = \sum_{n=-\infty}^{\infty} x(n)e^{-j\omega n} = -\sum_{n=-\infty}^{\infty} \alpha^n \mu(-n-1)e^{-j\omega n}$$

$$= -\sum_{n=-\infty}^{-1} \alpha^n e^{-j\omega n}.$$

Let $m = -n$, then

$$X(e^{j\omega}) = -\sum_{m=1}^{\infty} \alpha^{-m} e^{j\omega m} = -\alpha^{-1} e^{j\omega} \sum_{m=0}^{\infty} \left(\alpha^{-1}e^{j\omega}\right)^m$$

$$= -\alpha^{-1} e^{j\omega} \frac{1}{1 - \alpha^{-1}e^{j\omega}} \quad \text{for } |\alpha^{-1}e^{j\omega}| < 1$$

$$= \frac{1}{1 - \alpha e^{-j\omega}} \quad \text{for } |\alpha| > 1.$$

The rest of the solution is identical to Example 2.3. The difference between the two solutions is simply due to the regions of convergence being different. The plots of the magnitude and phase response for are shown in Figure 2.2.

The program for plotting the responses is included in the appendix as Program 2.3.

Example 2.4. Determine the DTFT of a causal exponential sequence $x(n) = \alpha^n \mu(n)$ for $\alpha = 1$. Notice that the exponential function reduces to a unit step function. Substituting $\alpha = 1$ in the result of Example 2.2 makes the geometric series nonconvergent. The solution will be obtained by taking the limit as α approaches 1.

Solution

$$X(e^{j\omega}) = \sum_{n=-\infty}^{\infty} x(n)e^{-j\omega n} = \sum_{n=-\infty}^{\infty} \alpha^n \mu(n)e^{-j\omega n} = \sum_{n=0}^{\infty} \alpha^n e^{-j\omega n}$$

$$= \sum_{n=0}^{\infty} (\alpha e^{-j\omega})^n = \frac{1}{1 - \alpha e^{-j\omega}} \quad \text{since } |\alpha e^{-j\omega}| < 1.$$

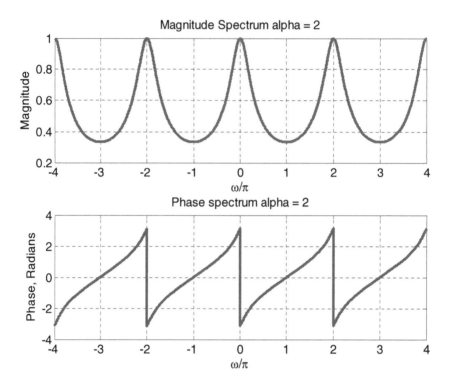

Fig. 2.2 Plot for the magnitude and phase response for $x(n) = \alpha^n \mu(n)$ for $\alpha > 1$.

For $x(n) = \mu(n) = \lim_{\alpha \to 1} \alpha^n \mu(n)$ and therefore

$$\text{DTFT} \left(\mu(n) \right) = \lim_{\alpha \to 1} \frac{1}{1 - \alpha e^{-j\omega}} = \frac{1}{1 - e^{-j\omega}} = e^{j\omega/2} \frac{1}{\sin(\omega/2)}.$$

The plot is shown in Figure 2.3 and the program is in the appendix as Program 2.4.

There are two important properties of the DTFT that needs to be highlighted.

(i) Although the time domain sequence is discrete the magnitude and phase frequency spectrum are continuous. For DSP processing this property poses a problem as a continuous frequency spectrum has an infinite number of sample points and would require an infinite memory capacity to store and an infinite time period to process.

(ii) As a result of sampling the DTFT is periodic. This can be verified analytically as follows:

Let $X(e^{j\omega})$ be the spectrum of a discrete-time signal.

If the spectrum is shifted by $2\pi k$ we get $X(e^{j\omega+j2\pi k}) = X(e^{j\omega})$ since $e^{j2\pi k} = 1$.

This implies that the spectrum of a digital signal is periodic with period 2π.

2.2 The Inverse Discrete-Time Fourier Transform

One of the properties of the DTFT which has been highlighted above is that it is periodic. This implies that it can be expressed as a Fourier series. In fact the DTFT expression of Equation (2.3) is a Fourier series

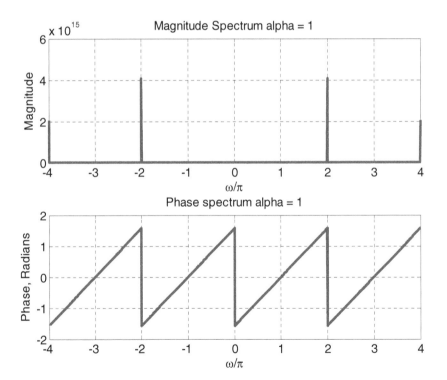

Fig. 2.3 Plot for the magnitude and phase response for $x(n) = \alpha^n \mu(n)$ for $\alpha = 1$.

Table 2.1 Some common DTFTs.

Sequence	DTFT	Condition
$\delta(n)$	1	
$\mu(n)$	$\dfrac{1}{1 - e^{-j\omega}} + \displaystyle\sum_{k=-\infty}^{\infty} 2\pi\delta(\omega + 2\pi k)$	
$\alpha^n \mu(n)$	$\dfrac{1}{1 - \alpha e^{-j\omega}}$	$\|\alpha\| < 1$
$-\alpha^n \mu(-n-1)$	$\dfrac{1}{1 - \alpha e^{-j\omega}}$	$\|\alpha\| > 1$
$ce^{-\alpha n}\mu(n)$	$\dfrac{c}{1 - e^{-\alpha}e^{-j\omega}}$	$e^{-\alpha} < 1$

with $x(n)$ representing the Fourier coefficients. Thus in order to find the inverse DTFT we can simply find the Fourier coefficients $x(n)$ of the Fourier series of $X(e^{j\omega})$. The expression for Fourier coefficients is given by

$$x(n) = \frac{1}{2\pi} \int_{-\pi}^{\pi} X(e^{j\omega}) e^{j\omega n} d\omega. \tag{2.11}$$

Example 2.5. The DTFT of a sequence $x(n)$ is given by $X(e^{j\omega}) = \sum_{k=-\infty}^{\infty} 2\pi\delta(\omega - \omega_0 + 2\pi k)$. Determine $x(n)$.

Solution

$$x(n) = \frac{1}{2\pi} \int_{-\pi}^{\pi} X(e^{j\omega}) e^{j\omega n} d\omega = \frac{1}{2\pi} \int_{-\pi}^{\pi} \sum_{k=-\infty}^{\infty} 2\pi\delta(\omega - \omega_0 + 2\pi k) e^{j\omega n} d\omega.$$

Excluding samples outside the range $-\pi \le \omega \le +\pi$ the integral reduces to

$$= \frac{1}{2\pi} \int_{-\pi}^{\pi} 2\pi\delta(\omega - \omega_0) e^{j\omega n} d\omega = e^{j\omega_0 n}.$$

Example 2.6. The DTFT of a sequence $x(n)$ is given by $X(e^{j\omega}) = \frac{1 - e^{-j\omega(N+1)}}{1 - e^{-j\omega}}$. Determine $x(n)$.

Solution

$$\frac{1 - e^{-j\omega(N+1)}}{1 - e^{-j\omega}} = \sum_{n=0}^{N} 1^{-n} e^{-j\omega n}.$$

This implies that $x(n) = 1$ for $0 \le n \le N$.

2.3 Properties of the Discrete-Time Fourier Transform

Some selected properties are verified here

(i) The Linearity Property: The discrete-time Fourier transform is a linear transformation. This can be proved by showing that the DTFT of a linear combination of two sequences is a linear combination of their DTFTs.

Proof. Let the sequences $x_1(n)$ and $x_2(n)$ have the DTFTs $X_1(e^{j\omega})$ and $X_2(e^{j\omega})$, respectively. Then

$$X_1(e^{j\omega}) = \sum_{n=-\infty}^{\infty} x_1(n)e^{-j\omega n} \quad \text{and}$$

$$X_2(e^{j\omega}) = \sum_{n=-\infty}^{\infty} x_2(n)e^{-j\omega n}.$$

A linear combination of $x_1(n)$ and $x_2(n)$ is given by $y(n) = \alpha x_1(n) + \beta x_2(n)$, where α and β are arbitrary constants. The DTFT of $y(n)$ is given by

$$Y(e^{j\omega}) = \sum_{n=-\infty}^{\infty} y(n)e^{-j\omega n} = \sum_{n=-\infty}^{\infty} (\alpha x_1(n) + \beta x_2(n))e^{-j\omega n}$$

$$= \sum_{n=-\infty}^{\infty} \alpha x_1(n)e^{-j\omega n} + \sum_{n=-\infty}^{\infty} \beta x_2(n)e^{-j\omega n}$$

$$= \alpha \sum_{n=-\infty}^{\infty} x_1(n)e^{-j\omega n} + \beta \sum_{n=-\infty}^{\infty} x_2(n)e^{-j\omega n}$$

$$= \alpha X_1(e^{j\omega}) + \beta X_2(e^{j\omega}),$$

which is a linear combination of the DTFTs of $x_1(n)$ and $x_2(n)$. Therefore the DTFT is a linear operator. $\square$

(ii) Time Shifting Property: If a sequence $x(n)$ with a DTFT $X(e^{j\omega})$ is shifted in time by n_0, to obtain a new sequence $x_1(n) = x(n - n_0)$, then the DTFT of the new sequence is given by $X(e^{j\omega}) = e^{-j\omega n_0}X(e^{j\omega})$.

Proof. By definition $X(e^{j\omega}) = \sum_{n=-\infty}^{\infty} x(n)e^{-j\omega n}$ and also

$$X_1(e^{j\omega}) = \sum_{n=-\infty}^{\infty} x_1(n)e^{-j\omega n} = \sum_{n=-\infty}^{\infty} x(n - n_0)e^{-j\omega n}.$$

Let $m = n - n_0$. Then

$$X_1(e^{j\omega}) = \sum_{n=-\infty}^{\infty} x(m)e^{-j\omega(m+n_0)}$$

$$= e^{-j\omega n_0} \sum_{n=-\infty}^{\infty} x(m)e^{-j\omega m} = e^{-j\omega n_0} X(e^{j\omega}). \quad \square$$

(iii) **The Differentiation Property:** If a sequence $x(n)$ has a DTFT given by $X(e^{j\omega})$ then the sequence $nx(n)$ has a DTFT given by $j\frac{dX(e^{j\omega})}{d\omega}$.

Proof. By definition $X(e^{j\omega}) = \sum_{n=-\infty}^{\infty} x(n)e^{-j\omega n}$. Differentiating with respect to ω gives $\frac{dX(e^{j\omega})}{d\omega} = \sum_{n=-\infty}^{\infty} -jnx(n)e^{-j\omega n}$, which can be written as $j\frac{dX(e^{j\omega})}{d\omega} = \sum_{n=-\infty}^{\infty} nx(n)e^{-j\omega n}$. This shows that $j\frac{dX(e^{j\omega})}{d\omega}$ is DTFT of the sequence $nx(n)$. $\quad \square$

(iv) **The Convolution Property:** If a sequence $x_1(n)$ has a DTFT $X_1(e^{j\omega})$ and a sequence $x_2(n)$ has a DTFT $X_2(e^{j\omega})$ then the DTFT of the sequence $x_1(n) \otimes x_2(n)$ (i.e., convolution of $x_1(n)$ and $x_2(n)$) is given by $X_1(e^{j\omega})X_2(e^{j\omega})$.

Proof. Let

$$X_1(e^{j\omega}) = \sum_{n=-\infty}^{\infty} x_1(n)e^{-j\omega n} \quad \text{and}$$

$$X_2(e^{j\omega}) = \sum_{n=-\infty}^{\infty} x_2(n)e^{-j\omega n}$$

$$X_3(e^{j\omega}) = \sum_{n=-\infty}^{\infty} x_1(n) \otimes x_2(n)e^{-j\omega n}$$

$$= \sum_{n=-\infty}^{\infty} \sum_{k=-\infty}^{\infty} x_1(n-k)x_2(k)e^{-j\omega n}.$$

Changing the order of summation, we get

$$X_3(e^{j\omega}) = \sum_{k=-\infty}^{\infty} \sum_{n=-\infty}^{\infty} x_1(n-k)x_2(k)e^{-j\omega n}$$

$$= \sum_{k=-\infty}^{\infty} x_2(k) \sum_{n=-\infty}^{\infty} x_1(n-k)e^{-j\omega n}.$$

Substituting $m = n - k$ we get

$$X_3(e^{j\omega}) = \sum_{k=-\infty}^{\infty} x_2(k) \sum_{n=-\infty}^{\infty} x_1(m)e^{-j\omega(m+k)}$$

$$= \sum_{k=-\infty}^{\infty} x_2(k)e^{-j\omega k} \sum_{n=-\infty}^{\infty} x_1(m)e^{-j\omega m}$$

$$= X_2(e^{j\omega})X_1(e^{j\omega}). \qquad \square$$

(v) Modulation Property: If a sequence $x_1(n)$ has a DTFT $X_1(e^{j\omega})$ and a sequence $x_2(n)$ has a DTFT $X_2(e^{j\omega})$ then the DTFT of the sequence $x_1(n) \cdot x_2(n)$ (i.e., product of $x_1(n)$ and $x_2(n)$) is given by $\frac{1}{2\pi} \int_{-\pi}^{\pi} X_1(e^{j\varphi})X_2^*(e^{j(\omega-\varphi)})d\varphi$.

Proof. Let

$$Y(e^{j\omega}) = \frac{1}{2\pi} \int_{-\pi}^{\pi} X_1(e^{j\varphi})X_2^*(e^{j(\omega-\varphi)})d\varphi.$$

Then

$$y(n) = \frac{1}{2\pi} \int_{\omega=-\pi}^{\pi} Y(e^{j\omega})e^{j\omega n} d\omega$$

$$= \frac{1}{2\pi} \int_{\omega=-\pi}^{\pi} \frac{1}{2\pi} \int_{\varphi=-\pi}^{\pi} X_1(e^{j\varphi})X_2^*(e^{j(\omega-\varphi)})d\varphi e^{j\omega n} d\omega.$$

Changing the order of integration and rearranging, we get

$$y(n) = \frac{1}{2\pi} \int_{\varphi=-\pi}^{\pi} X_1(e^{j\varphi}) \frac{1}{2\pi} \int_{\omega=-\pi}^{\pi} X_2^*(e^{j(\omega-\varphi)})e^{j\omega n} d\varphi d\omega.$$

Let $\theta = \omega - \varphi$ and therefore $\omega = \theta + \varphi$

$$y(n) = \frac{1}{2\pi} \int_{\varphi=-\pi}^{\pi} X_1(e^{j\varphi})$$

$$\times \frac{1}{2\pi} \int_{\theta=-\pi-\varphi}^{\pi-\varphi} X_2(e^{j\theta}) e^{-j(\theta+\varphi)n} d\varphi (d\theta + d\varphi)$$

$$= \frac{1}{2\pi} \int_{\varphi=-\pi}^{\pi} X_1(e^{j\varphi}) e^{-j\varphi n} d\varphi \frac{1}{2\pi} \int_{\theta=-\pi-\varphi}^{\pi-\varphi} X_2(e^{j\theta}) e^{-j\theta n} d\theta$$

$$= x_1(n) \cdot x_2(n).$$ $\quad\square$

(vi) Parseval's Relation: If a sequence $x_1(n)$ has a DTFT $X_1(e^{j\omega})$ and a sequence $x_2(n)$ has a DTFT $X_2(e^{j\omega})$ then Parseval's relation gives the relation between the cross-product terms in the time domain and the frequency domain

$$\sum_{n=-\infty}^{\infty} x_1(n)x_2^*(n) = \frac{1}{2\pi} \int_{-\pi}^{\pi} X_1(e^{j\omega})X_2^*(e^{j\omega})d\omega. \quad (2.12)$$

When $x_1(n) = x_2(n)$ the relation reduces to the total energy computed in the time domain in relation to the total energy obtained from the integral of the energy spectral density in the frequency domain

$$\sum_{n=-\infty}^{\infty} |x_1(n)|^2 = \frac{1}{2\pi} \int_{-\pi}^{\pi} |X_1(e^{j\omega})|^2 d\omega. \quad (2.13)$$

2.4   Linear Convolution

In LTI discrete-time systems the impulse response can completely characterize the system. Given the impulse response the output of the LTI discrete-time system can be computed for any arbitrary input. Suppose the impulse response of an LTI system to an impulse $\delta(n)$ is given by $h(n)$. The impulse response to a delayed version of the impulse, $\delta(n - k)$ will be given by $h(n - k)$ since the system is time invariant. In practice the input signal $x(n)$ is expressed as a sum of weighted and delayed impulses as in Equation (2.14).

$$x(n) = \sum_{k=-\infty}^{\infty} x(k)\delta(n - k), \quad (2.14)$$

Table 2.2 Summary of properties of DTFT properties.

Property	Sequence $x_1(n)$, $x_2(n)$	DTFT $X_1(e^{j\omega})$, $X_2(e^{j\omega})$
Linearity	$\alpha x_1(n) + \beta x_2(n)$	$\alpha X_1(e^{j\omega}) + \beta X_2(e^{j\omega})$
Time-shifting	$x_1(n - n_0)$	$e^{-j\omega n_0} X(e^{j\omega})$
Frequency-shifting	$e^{j\omega_0} x_1(n)$	$X(e^{j(\omega - \omega_0)})$
Frequency differentiation	$n x_1(n)$	$j\dfrac{dX(e^{j\omega})}{d\omega}$
Convolution	$x_1(n) \otimes x_2(n)$	$X_1(e^{j\omega})X_2(e^{j\omega})$
Modulation	$x_1(n)x_2(n)$	$\dfrac{1}{2\pi}\displaystyle\int_{-\pi}^{\pi} X_1(e^{j\varphi})X_2^*(e^{j(\omega-\varphi)})d\varphi$
Parseval's relation	$\displaystyle\sum_{n=-\infty}^{\infty} x_1(n)x_2^*(n)$	$\dfrac{1}{2\pi}\displaystyle\int_{-\pi}^{\pi} X_1(e^{j\omega})X_2^*(e^{j\omega})d\omega$

where $x(k)$ is the weight and $\delta(n - k)$ gives location of the sample in terms of the time instance the sample appears. Due to the linearity property of the LTI system we can compute the response due to each sample and add them up to obtain the output of the system as shown in Equation (2.15).

$$y(n) = \sum_{k=-\infty}^{\infty} x(k)h(n - k). \tag{2.15}$$

Equation (2.15) is what is referred to as the convolution sum. The notation and the alternative form are given by Equation (2.16).

$$y(n) = x(n) \otimes h(n) = \sum_{k=-\infty}^{\infty} x(k)h(n - k) = \sum_{k=-\infty}^{\infty} h(k)x(n - k). \tag{2.16}$$

Equation (2.9) directs us to compute the output using the following procedure for each output sample starting with $n = 0$:

(i) Obtain the time reversal of the impulse response $h(-k)$.
(ii) For each value of n shift $h(-k)$, n times to obtain $h(n - k)$.
(iii) Multiply the overlapping samples of $x(k)$ and $h(n - k)$.
(iv) The sum of products $x(k)h(n - k)$ of the overlapping samples gives $y(n)$ for the specific value of n.
(iv) Continue with the process until there are no more overlapping samples.

The convolution process is feasible when the impulse response is of finite length and the input signal is also of finite length. In such a situation the output sequence will also be of finite length. It is also possible to compute an output when the impulse response is of finite length and the input is an endless stream of data. This is because in order to compute each sample of the output there is a finite sum of products to add. The convolution sum fails to compute the output when both the input signal and the impulse response are of infinite duration. Whenever the impulse response is of infinite duration the practice is to use linear constant coefficients difference equations to compute the output.

2.4.1   Graphical Implementation of Linear Convolution

In the next example we show how the convolution process is achieved graphically.

Example 2.7. An input sequence $x(n)$ and an impulse response $h(n)$ to an LTI system are shown in Figure 2.4. By means of convolution shown graphically in Figure 2.5 obtain the output of the system.

Solution

The output sequence is given by the convolution sum as

$$y(n) = \sum_{k=-\infty}^{\infty} x(k)h(n-k).$$

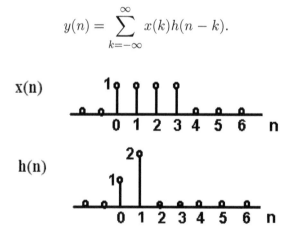

Fig. 2.4 Input sequence $x(n)$ and impulse response $h(n)$.

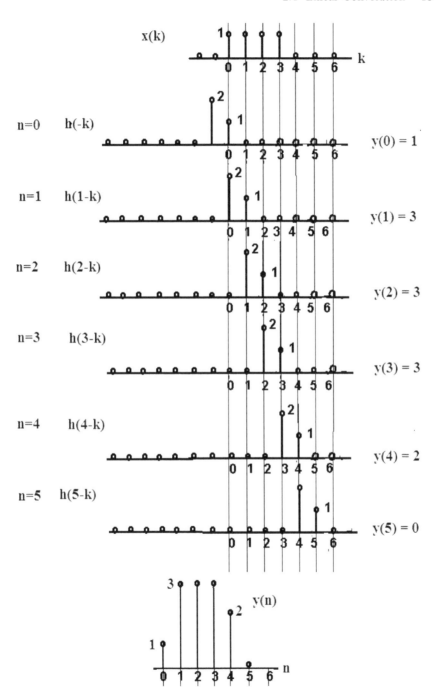

Fig. 2.5 Graphical linear convolution.

2.4.2   Implementation of Linear Convolution Using DTFTs

It is possible to implement linear convolution through the use of DTFTs. If two sequences $x_1(n)$ and $x_2(n)$ are to be convolved to obtain the result $y(n) = x_1(n) \otimes x_2(n)$, we can achieve the same result by taking the inverse DTFT of the product of their DTFTs as shown in Figure 2.6.

2.5   MATLAB Plots of DTFTs

Using some functions in MATLAB a program to plot DTFT or frequency spectrum of any rational function can be written (see Figure 2.7). The key MATLAB functions are freqs and freqz which are used for analog and digital frequency spectrum, respectively. In order to get the syntax and other variations of the syntax one must type "Help freqs" or "Help freqz" in the MATLAB command window.

In our application freqz can be used to calculate the vector h representing the magnitude and phase of the transfer function at specific frequency ω of a rational transfer function given the coefficients. The syntax is given as

$$h = freqz(b, a, \omega), \tag{2.17}$$

where b and a are row matrices representing the numerator and denominator coefficients, respectively.

A program is developed below to show how the MATLAB functions are applied to plot the frequency spectrum of a rational function given

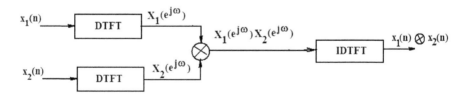

Fig. 2.6 Implementation of linear convolution using DTFTs.

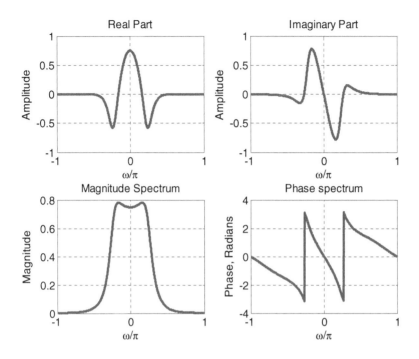

Fig. 2.7 Frequency spectrum from MATLAB functions.

in Equation (2.18)

$$H(e^{j\omega}) = \frac{0.0098 + 0.0393e^{-j\omega} + 0.0590e^{-j2\omega} + 0.0098e^{-j3\omega}}{1 - 1.9908e^{-j\omega} + 1.7650e^{-j2\omega} - 0.7403e^{-j3\omega} + 0.1235e^{-j4\omega}}.$$

(2.18)

Program 2.1

```
% Plotting of the Frequency response of a rational function
% Enter the desired length of the DFT
k = input('Enter the number of frequency points = ');
% Enter the numerator and denominator coefficients
num = input('Enter the numerator coefficients = ');
den = input('Enter the denominator coefficients = ');
% Compute the frequency response
w = -pi:pi/k:pi;
h = freqz(num, den, w);
```

```
subplot(2,2,1);
plot(w/pi, real(h)); grid
title('Real Part');
xlabel('\omega/\pi'); ylabel('Amplitude');
subplot(2,2,2);
plot(w/pi, imag(h)); grid
title('Imaginary Part');
xlabel('\omega/\pi'); ylabel('Amplitude');
subplot(2,2,3);
plot(w/pi, abs(h)); grid
title('Magnitude Spectrum');
xlabel('\omega/\pi'); ylabel('Magnitude');
subplot(2,2,4);
plot(w/pi, angle(h)); grid
title('Phase spectrum');
xlabel('\omega/\pi'); ylabel('Phase, Radians');
```

The Plots of $H(e^{j\omega})$

2.6   Problems

2.6.1 Determine the DTFTs of the following sequences:

(i) $w(n) = \beta^n \mu(n-2)$ for $\beta < 1$,

(ii) $x(n) = \begin{cases} \beta^{|n|} & |n| \le L \\ 0 & \text{otherwise} \end{cases}$, and

(iii) $y(n) = \beta^n \mu(-n-4)$ for $\beta > 1$.

2.6.2 If $x(n)$ is a real sequence with a DTFT $X(e^{j\omega})$ determine the DTFT of $x(-n)$ in terms of $X(e^{j\omega})$.

2.6.3 If $X(e^{j\omega})$ is the DFFT of a real sequence $x(n)$, determine the inverse DFTS of

(i) $X_{re}(e^{j\omega})$ (ii) $X_{im}(e^{j\omega})$ in terms of $x(n)$.

2.6.4 For a sequence $x(n) = [121013]$ for $-1 \le n \le 4$, without computing the DTFT evaluate

(i) the DC component of the frequency,

(ii) the integral $\int_{-\pi}^{\pi} X(e^{j\omega})d\omega$, and

(iii) the integral $\int_{-\pi}^{\pi} |X(e^{j\omega})|^2 d\omega$.

2.6.5 A finite length sequence is given by $x(n) = [1\ 2\ 3\ 2\ 1]$ for $0 \le n \le 4$ has DTFT $X(e^{j\omega})$. Show that this sequence has a linear phase.

2.6.6 The output of an FIR filter is given by the following equation:
$y(n) = x(n) - 2x(n-1) + 3x(n-2) + 2x(n-3) - x(n-4)$. Show that the filter has linear phase.

2.6.7 If the complex sequence $x(n)$ has a DTFT $X(e^{j\omega})$ prove the following symmetry properties

(i) the DTFT of $x(-n)$ is $X(e^{-j\omega})$,

(ii) the DTFT of $x^*(-n)$ is $X(e^{j\omega})$,

(iii) the DTFT of $\text{Re}\{x(n)\}$ is $\frac{1}{2}\{X(e^{j\omega}) + X^*(e^{-j\omega})\}$,

(iv) the DTFT of $j\text{Im}\{x(n)\}$ is $\frac{1}{2}\{X(e^{j\omega}) - X^*(e^{-j\omega})\}$.

2.6.8 If the real sequence $x(n)$ has a DTFT $X(e^{j\omega})$ prove the following symmetry properties

(i) $X(e^{j\omega}) = X^*(e^{-j\omega})$ (ii) $X_{\text{re}}(e^{j\omega}) = X_{\text{re}}(e^{-j\omega})$ (iii) $X_{\text{im}}(e^{j\omega}) = -X_{\text{im}}(e^{-j\omega})$.

2.6.9 An IIR filter frequency response is given by

$$H(e^{j\omega}) = 0.0124 \left(\frac{1 + e^{-j\omega}}{1 - 0.6386e^{-j\omega}} \right)$$

$$\times \left(\frac{1 + 1.5975e^{-j\omega} + e^{-j2\omega}}{1 - 0.9621e^{-j\omega} + 0.5708e^{-j2\omega}} \right)$$

$$\times \left(\frac{1 + 1.12724e^{-j\omega} + e^{-j2\omega}}{1 - 0.5811e^{-j\omega} + 0.8547e^{-j2\omega}} \right).$$

Using MATLAB to plot the following frequency responses

(i) magnitude response,

(ii) phase response,

(iii) of the real part of $H(e^{j\omega})$, and

(iv) the imaginary part of $H(e^{j\omega})$.

3

The Transform Domain Analysis: The Discrete Fourier Transform

3.1 The Discrete Fourier Transform

In Chapter 2 we observed that the DTFT $X(e^{j\omega})$ of a sequence $x(n)$ is continuous with respect to frequency. The continuous spectrum is made up of an infinite number of samples that requires infinite memory to store and infinite time to process. It is not feasible to process such a signal in a DSP processor. The solution to this problem is to process a finite number of samples taken from $X(e^{j\omega})$. The number of samples taken is normally made equal to the number of samples of the sequence $x(n)$. Such samples form what is referred to as the Discrete Fourier Transform (DFT) which is defined only for finite length sequences as follows:

The DFT of a finite length sequence $x(n)$, $0 \leq n \leq N - 1$ is defined as

$$X(k) = \sum_{n=0}^{N-1} x(n)e^{-j2\pi kn/N} \quad \text{for } 0 \leq k \leq N - 1. \tag{3.1}$$

It must be noted here that whereas n is a time index, k is a frequency index. $X(k)$s are referred to as the DFT coefficients. The expression for DFT is very similar to that of the DTFT with $e^{j\omega}$ being replaced by the frequency samples $e^{j2\pi k/N}$ and the range limited to $0 \leq k \leq N - 1$. The process of finding the frequency spectrum of a signal using the DFT is referred to as the analysis process and is represented by Equation (3.1). On the other hand, knowing the spectrum it is possible to find the original sequence. The process of finding the original sequence from the spectrum here represented by DFT coefficients is referred to as the synthesis process and is represented by Equation (3.1) which is an expression of the inverse DFT (IDFT).

$$x(n) = \frac{1}{N} \sum_{k=0}^{N-1} X(k)e^{j2\pi kn/N} \quad \text{for } 0 \leq n \leq N - 1. \tag{3.2}$$

In order to simplify the representation we define a twiddle factor as $W_N = e^{-j\frac{2\pi}{N}}$. Substituting the twiddle factor into Equations (3.1) and (3.2) we get more compact expressions for the DFT

$$X(k) = \sum_{n=0}^{N-1} x(n)W_N^{kn} \quad \text{for } 0 \leq k \leq N - 1, \tag{3.3}$$

and the IDFT is given by

$$x(n) = \frac{1}{N} \sum_{k=0}^{N-1} X(k)W_N^{-kn} \quad \text{for } 0 \leq n \leq N - 1. \tag{3.4}$$

Example 3.1. Determine the DTFT of a length-4 sequence given by $x(n) = \{(0\ 1\ 0\ 0)\}$.

Solution

$$X(k) = \sum_{n=0}^{N-1} x(n)W_N^{kn} \quad \text{for } 0 \leq k \leq N - 1$$

$$X(k) = 0W_4^0 + W_4^k + 0W_4^{2k} + 0W_4^{3k} = W_4^k \quad \text{for } k = 0,1,2,3$$

$X(0) = 1$, $X(1) = -j$, $X(2) = -1$, $X(3) = j$ or as a row matrix $X(k) = [1, -j, -1,\ j]$.

Example 3.2. Determine the DTFT of a length-4 sequence given by $x(n) = \{(1,1,1,1)\}$.

Solution

$$X(k) = \sum_{n=0}^{N-1} x(n)W_N^{kn} \quad \text{for } 0 \leq k \leq N - 1$$

$$X(k) = \quad \text{for } k = 0,1,2,3$$

$$X(0) = 1 + 1 + 1 + 1 = 4$$

$$X(1) = W_4^0 + W_4^1 + W_4^2 + W_4^3 = 1 - j - 1 + j = 0$$

$$X(2) = W_4^0 + W_4^2 + W_4^4 + W_4^6 = 1 - 1 + 1 - 1 = 0$$

$$X(3) = W_4^0 + W_4^3 + W_4^6 + W_4^9 = 1 - j - 1 + j = 0.$$

As a row matrix as a row matrix $X(k) = [4, 0, 0, 0]$.

Sometimes writing Equations (3.1) and (3.2) in matrix notation simplifies the computation of the DFT or its inverse. Equation (3.1) can be written in matrix notation as follows:

$$\begin{bmatrix} X(0) \\ X(1) \\ \vdots \\ X(N-1) \end{bmatrix} = \begin{bmatrix} W_N^0 & W_N^0 & W_N^0 & \cdots & W_N^0 \\ W_N^0 & W_N^1 & W_N^2 & \cdots & W_N^{(N-1)} \\ W_N^0 & W_N^2 & W_N^4 & \cdots & W_N^{2(N-1)} \\ \vdots & \vdots & \vdots & \vdots & \vdots \\ W_N^0 & W_N^{(N-1)1} & W_N^{(N-1)2} & \cdots & W_N^{(N-1)(N-1)} \end{bmatrix}$$

$$\times \begin{bmatrix} x(0) \\ x(1) \\ x(2) \\ \vdots \\ x(N-1) \end{bmatrix}, \tag{3.5}$$

which can be written in vector form as

$$\overline{X(k)} = \overline{W_N} \overline{x(n)}, \tag{3.6}$$

where

$$\overline{X(k)} = [X(0), X(1), \dots, X(N-1)]^T,$$
$$\overline{x(n)} = [x(0), x(1), \dots, (x(N-1))]^T,$$

and

$$\overline{W_N} = \begin{bmatrix} W_N^0 & W_N^0 & W_N^0 & \cdots & W_N^0 \\ W_N^0 & W_N^1 & W_N^2 & \cdots & W_N^{(N-1)} \\ W_N^0 & W_N^2 & W_N^4 & \cdots & W_N^{2(N-1)} \\ \vdots & \vdots & \vdots & \vdots & \vdots \\ W_N^0 & W_N^{(N-1)1} & W_N^{(N-1)2} & \cdots & W_N^{(N-1)(N-1)} \end{bmatrix}.$$

We can also write Equation (3.4) as follows:

$$\overline{x(n)} = \frac{1}{N}\overline{W_N}^{-1}\overline{X(k)}, \tag{3.7}$$

where $\overline{W_N}^{-1} = \overline{(W_N)}^*$.

Example 3.3. We repeat part of Example 3.2 using matrices and go further to obtain the IDFT also using matrices. The sequence used is a length-4 sequence given by $x(n) = \{(1\ 1\ 1\ 1)\}$.

Solution

$$\begin{bmatrix} X(0) \\ X(1) \\ X(2) \\ X(3) \end{bmatrix} = \begin{bmatrix} W_N^0 & W_N^0 & W_N^0 & W_N^0 \\ W_N^0 & W_N^1 & W_N^2 & W_N^3 \\ W_N^0 & W_N^2 & W_N^4 & W_N^6 \\ W_N^0 & W_N^3 & W_N^6 & W_N^9 \end{bmatrix} \begin{bmatrix} x(0) \\ x(1) \\ x(2) \\ x(3) \end{bmatrix}.$$

To simplify we must use the property of the twiddle factor that for $n > N, W_N^n = \langle W_N^n \rangle_N$, where the operator $\langle\rangle_N$ implies modulo N operation.

Therefore   substituting $W_4^4 = W_4^0, W_4^6 = W_4^2, W_4^9 = W_4^1$ in   the matrix equation we obtain

$$\begin{bmatrix} X(0) \\ X(1) \\ X(2) \\ X(3) \end{bmatrix} = \begin{bmatrix} W_4^0 & W_4^0 & W_4^0 & W_4^0 \\ W_4^0 & W_4^1 & W_4^2 & W_4^3 \\ W_4^0 & W_4^2 & W_4^0 & W_4^2 \\ W_4^0 & W_4^3 & W_4^2 & W_4^1 \end{bmatrix} \begin{bmatrix} 1 \\ 1 \\ 1 \\ 1 \end{bmatrix}.$$

Evaluating $W_4^0 = 1, W_4^1 = e^{-i2\pi/4} = -j, W_4^2 = (-j)^2 = -1, W_4^3 = (-j)(-1) = j$.

$$\begin{bmatrix} X(0) \\ X(1) \\ X(2) \\ X(3) \end{bmatrix} = \begin{bmatrix} 1 & 1 & 1 & 1 \\ 1 & -j & -1 & j \\ 1 & -1 & 1 & -1 \\ 1 & j & -1 & -j \end{bmatrix} \begin{bmatrix} 1 \\ 1 \\ 1 \\ 1 \end{bmatrix} = \begin{bmatrix} 4 \\ 0 \\ 0 \\ 0 \end{bmatrix},$$

Therefore $X(k) = [4, 0, 0, 0]$, which is the same answer as we got in Example 3.2.

In order to find the IDFT we need to find $\overline{W_N}^{-1} = \overline{(W_N)}^{*}$. This is obtained by conjugating each element of $\overline{W_N}$ as follows:

$$\begin{bmatrix} x(0) \\ x(1) \\ x(2) \\ x(3) \end{bmatrix} = \frac{1}{4} \begin{bmatrix} 1 & 1 & 1 & 1 \\ 1 & +j & -1 & -j \\ 1 & -1 & 1 & -1 \\ 1 & -j & -1 & +j \end{bmatrix} \begin{bmatrix} 4 \\ 0 \\ 0 \\ 0 \end{bmatrix} = \frac{1}{4} \begin{bmatrix} 4 \\ 4 \\ 4 \\ 4 \end{bmatrix} = \begin{bmatrix} 1 \\ 1 \\ 1 \\ 1 \end{bmatrix}.$$

3.2 MATLAB Plots of DFTs

Using some functions in MATLAB a program to compute and plot the DFT of any sequence can be written. The key MATLAB functions are $FFT(x)$ and $IFFT(X)$ which make use of the efficient Fast Fourier transform in the computation. An alternative syntax uses $FFT(X, N)$ and $IFFT(x, N)$ which specifies the length of the DFT. If x has a length less than N then x is padded with zeros.

3.2.1 MATLAB Program for Plotting DFT

Program 3.1

```
% A MATLAB programme to plot the input sequence,
   compute and plot the
% magnitude and phase of the DFT. The length of
   the sequence and the
% length of the desired DFT are entered as inputs
N = input('Length of sequence = ');
M = input('Length of DFT = ');
x = input('Enter the sequence as a row matrix x = ');
% Obtain the M-point DFT
X = fft(x, M);
% Plot the time-domain sequence
l = 0:1:N-1;
subplot(3,1,1)
stem(l,x);
title('Sequence in the Time Domain');
xlabel('Time Index n'); ylabel('Amplitude');
```

```
% Plot the Magnitude of the DFT samples
subplot(3,1,2)
k = 0:1:M-1;
stem(k, abs(X))
title('Magnitude of the DFT samples');
xlabel('Frequency Index k'); ylabel('Magnitude');
subplot(3,1,3)
stem(k, angle(X))
title('Phase of the DFT samples');
xlabel('Frequency Index k'); ylabel('Phase');
```

Example 3.4. Use of MATLAB functions to plot the DFT of a sequence in Program 3.1 (see Figure 3.1).

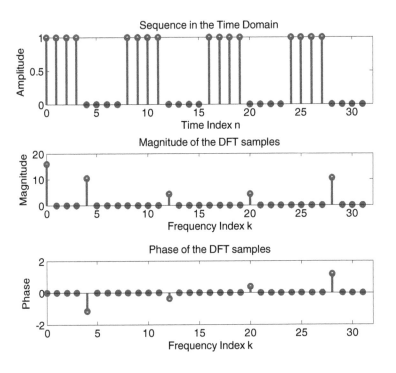

Fig. 3.1 A 32-point DFT of a 32-point sequence.

3.2.2   MATLAB Program for Plotting an IDFT

Program 3.2

```
% A MATLAB programme to plot a given DFT,
  compute and plot the real &
% imaginary part of its IDFT
% The length of the DFT and the desired IDFT
  are read in
K = input('Length of DFT K = ');
N = input('Length of IDFT N = ');
X = input('Enter the DFT as a row matrix X = ');
% Obtain the N-point IDFT
x = ifft(X, N);
% Plot the original DFT samples
k = 0:1:K-1;
subplot(3,1,1)
stem(k,X);
title('DFT Samples Provided');
xlabel('Frequency Index k'); ylabel('Amplitude');
% Plot the real part of the DFT samples
subplot(3,1,2)
n = 0:1:N-1;
stem(n, real(x))
title('Real part of x(n), the IDFT samples');
xlabel('Time Index n'); ylabel('Amplitude');
subplot(3,1,3)
stem(n, imag(x))
title('Imaginary part of x(n), the IDFT samples');
xlabel('Time Index n'); ylabel('Amplitude')
```

Example 3.5. Use of MATLAB functions to obtain the inverse DFT of a sequence (see Figure 3.2).

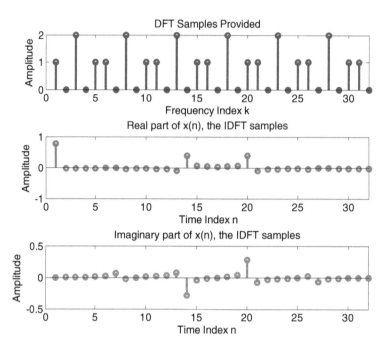

Fig. 3.2 A 32-point IDFT from a 32-point DFT.

3.2.3   MATLAB Program for Estimating the DTFT From the DFT

Program 3.3

```
% A MATLAB programme to plot an estimated spectrum from
   DFT samples
% of a 50 Hz square wave
% The programme input will be an analog input which
   will be plotted
% The analog square wave at 50 Hz will be sampled
   and plotted
% The DFT will be computed and plotted
% A DTFT will be estimated from the DFT
N = input('Length of input sequence = ');
M = input('Desired Length of the DFT = ');
% For example, generate a 50 Hz square wave:
```

```
t = 0:2/(50*1000):1/50;
subplot(5,1,1)
y = square(2*pi*50*t); plot(t,y);
% The analog signal\index{analog signal} is sampled
  to obtain a
digital signal\index{digital signal}
n = 0:N;
x = square(2*pi*n*1/50);
% Plot the time domain sequence
subplot(5,1,2)
stem(n,x);
title('Sequence in the Time Domain');
xlabel('Time Index n'); ylabel('Amplitude');
% Obtain and plot the magnitude and phase of the DFT
X = fft(x ,M);
% Plot the Magnitude of the DFT samples
subplot(5,1,3)
k = 0:1:M-1;
stem(k, abs(X))
title('Magnitude of the DFT samples');
xlabel('Frequency Index k'); ylabel('Magnitude');
subplot(5,1,4)
stem(k, angle(X))
title('Phase of the DFT samples');
xlabel('Frequency Index k'); ylabel('Phase');
{\%} Compute 512-point DFT x
XE = fft(x,512);
% Plot the frequency response
L = 0:511;
subplot(5,1,5)
plot(L/512,abs(XE));
hold
plot(k/M,abs(X),'o')
title('Estimation Spectrum of a Square Wave')
xlabel('Normalised Angular Frequency')
ylabel('magnitude')
```

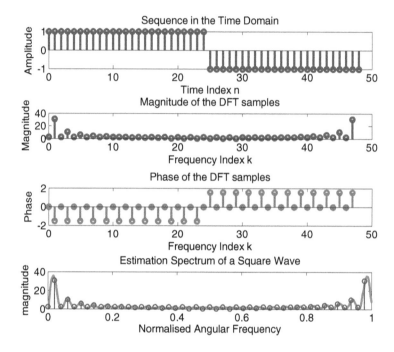

Fig. 3.3 Estimation of the spectrum of a discrete-time square wave.

Example 3.6. Starting with a discrete-time square waveform its DFT is computed and finally its DTFT or its spectrum is obtained. The last plot shows the DFT superimposed on to the DTFT (see Figure 3.3).

3.3   Discrete Fourier Transform Properties

Before we introduce properties of the DFT we will first introduce the concept of circular shift. Certain sequences are defined over a specified interval for instance in the range $0 \leq n \leq N - 1$ and they are not defined outside this interval. If a time shift is going to be made on a sequence $w(n)$ such as $w(n - n_0)$ the sequence will be extended into the range in which it is not defined. To prevent this from happening we define a different type of shift; a circular shift. This is similar to having N samples of the sequence $x(n)$ uniformly distributed on the circumference of a circle. Shifting a sequence n_0 samples to the right is the same as shifting the sequence n_0 samples in the counter-clockwise direction along the circumference with respect to a reference. This is equivalent to the $(N-1)$th (or last) sample before a shift moving to the 0th (first) sample after a shift on a linear axes. Mathematically this can be defined using a modulo arithmetic as follows:

$$w_c(n) = w(\langle n - n_0 \rangle_N),\qquad\qquad(3.8)$$

where $w(n)$ is the original sequence and the shifted sequence is $w(\langle n - n_0 \rangle_N)$. This is shown in Figure 3.4.

The properties of DFTs are similar to those of DTFTs and they are useful in the simplification of DFT computation and implementation. In this section, we have focused on the general properties which is depicted in Table 3.1. The remaining properties based on symmetry may be introduced in the exercises at the end of the chapter and some can be found in more advanced text books [3].

(i) Linearity Property: The discrete Fourier transform is a linear transformation. This can be proved by showing that the DFT of a linear combination of two sequences is a linear combination of their DFTs.

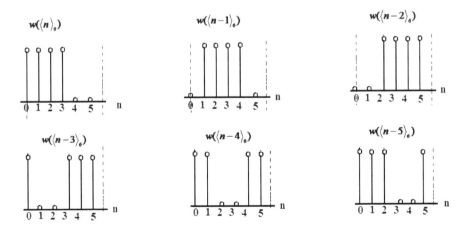

Fig. 3.4 Circular shift of a finite length sequence.

Table 3.1 Summary of the DFT properties.

Property	Sequence	DFT
Linearity	$\alpha x_1(n) + \beta x_2(n)$	$\alpha X_1(k) + \beta X_2(k)$
Circular shift	$x(\langle n - n_0 \rangle_N)$	$W_N^{kn_0} X(k)$
Frequency shift	$W_N^{-k_0 n} x(n)$	$X(\langle k - k_0 \rangle_N)$
Convolution	$x(n) \circledast h(n)$	$X(k)H(k)$
Modulation	$x_1(n)x_2(n)$	$\frac{1}{N} \sum_{m=0}^{N-1} X_1(l)X_2(k-l)$
Parseval's relation	$\sum_{n=0}^{N-1} \|x(n)\|^2$	$\frac{1}{N} \sum_{k=0}^{N-1} \|X(k)\|^2$

Proof. Let the sequences $x_1(n)$ and $x_2(n)$ have the DFTs $X_1(k)$ and $X_2(k)$, respectively. Then

$$X_1(k) = \sum_{n=0}^{N-1} x_1(n)W_N^{kn} \quad \text{for } 0 \le k \le N - 1, \quad \text{and}$$

$$X_2(k) = \sum_{n=0}^{N-1} x_2(n)W_N^{kn} \quad \text{for } 0 \le k \le N - 1.$$

A linear combination of $x_1(n)$ and $x_2(n)$ is given by $y(n) = \alpha x_1(n) + \beta x_2(n)$, where α and β are arbitrary constants. The DFT of $y(n)$ is given by

$$Y(k) = \sum_{n=0}^{N-!} y(n)W_N^{kn}$$

$$= \sum_{n=0}^{N-1} (\alpha x_1(n) + \beta x_2(n))W_N^{kn} \quad \text{for } 0 \le k \le N - 1$$

$$= \sum_{n=0}^{N-1} \alpha x_1(n)W_N^{kn} + \sum_{n=0}^{N-1} \beta x_2(n)W_N^{kn}$$

$$= \alpha \sum_{n=0}^{N-1} x_1(n)W_N^{kn} + \beta \sum_{n=0}^{N-1} x_1(n)W_N^{kn}$$

$$= \alpha X_1(k) + \beta X_2(k) \quad \text{for } 0 \le k \le N - 1,$$

which is a linear combination of the DFTs of $x_1(n)$ and $x_2(n)$. Therefore the DFT is a linear operator. $\square$

(ii) Circular Shift Property: If $x(n)$ has DFT $X(k)$ then the DFT of the circular shift of $x_1(n) = x(\langle n - n_0 \rangle_N)$ is given by $X_1(k) = W^{kn_0} X(k)$.

Proof.

$$X_1(k) = \sum_{n=0}^{N-1} x_1(n) W_N^{kn}$$

$$= \sum_{n=0}^{N-1} x(\langle n - n_0 \rangle_N) W_n^{kn} \quad \text{for } 0 \leq k \leq N - 1$$

Let $m = n - n_0$, then

$$X_1(k) = \sum_{m=0}^{N-1} x(\langle m \rangle_N) W_n^{k(m+n_0)}$$

$$= W_N^{kn_0} \sum_{m=0}^{N-1} x(\langle m \rangle_N) W_N^{km} \quad \text{for } 0 \leq k \leq N - 1$$

$$= W_N^{kn_0} X(k) \quad \text{for } 0 \leq k \leq N - 1. \qquad \square$$

(iii) Frequency Shift: If $x(n)$ has DFT $X(k)$ then the DFT of $x_2(n) = W_N^{-k_0 n} x(n)$ is given by $X_2(k) = X(\langle k - k_0 \rangle_N)$.

Proof.

$$X_2(k) = \sum_{n=0}^{N-1} x_2(n) W_N^{kn}$$

$$= \sum_{n=0}^{N-1} x(n) W_N^{-k_0 n} W_n^{kn} \quad \text{for } 0 \leq k \leq N - 1$$

$$= \sum_{n=0}^{N-1} x(n) W_N^{(k-k_0)} = X(\langle k - k_0 \rangle_N). \qquad \square$$

(iv) Circular Convolution: If $y(n)$ is a circular convolution of two sequences $x(n)$ and $h(n)$, i.e., $y(n) = \sum_{m=0}^{N-1} x(m) h(\langle n - m \rangle_N)$ then the DFT of $y(n)$ is given by $Y(k) = X(k) H(k)$.

Proof.

$$Y(k) = \sum_{n=0}^{N-1} y(n) W_N^{kn}$$

$$= \sum_{n=0}^{N-1} \sum_{m=0}^{N-1} x(m) h(\langle n - m \rangle_N) W_N^{kn} \quad \text{for } 0 \le k \le N - 1.$$

Substituting $l = \langle n - m \rangle_N$ and therefore $n = \langle l + m \rangle_N$ and changing the order of summation

$$Y(k) = \sum_{n=0}^{N-1} \sum_{m=0}^{N-1} x(m) h(l) W_N^{(l+m)k}$$

$$= \sum_{m=0}^{N-1} x(m) W_N^{km} \sum_{n=0}^{N-1} h(l) W_N^{kl}$$

$$= X(k) H(k) \quad \text{for } 0 \le k \le N - 1. \qquad \square$$

(v) Modulation: If $y(n)$ is the modulation product of two sequences $x_1(n)$ and $x_2(n)$, i.e., $y(n) = x_1(n) x_2(n)$, then the DFT of $y(n)$ is the frequency convolution of the DFT of $x_1(n)$ and the DFT of $x_2(n)$, i.e., $Y(k) = \frac{1}{N} \sum_{m=0}^{N-1} X_1(l) X_2(k - l)$.

Proof.

$$Y(k) = \sum_{n=0}^{N-1} y(n) W_N^{kn}$$

$$= \sum_{n=0}^{N-1} x_1(n) x_2(n) W_N^{kn} \quad \text{for } 0 \le k \le N - 1.$$

But

$$x_1(n) = \frac{1}{N} \sum_{l=0}^{N-1} X_1(l) W_N^{-nl} \quad \text{for } 0 \le n \le N - 1.$$

Substituting we get

$$Y(k) = \sum_{n=0}^{N-1} \left(\frac{1}{N} \sum_{l=0}^{N-1} X_1(l) W_N^{-nl} \right) x_2(n) W_N^{kn}.$$

Interchanging the order of summation

$$Y(k) = \frac{1}{N}\sum_{l=0}^{N-1}X_1(l)\sum_{n=0}^{N-1}x_2(n)W_N^{-nl}W_N^{kn}$$

$$= \frac{1}{N}\sum_{l=0}^{N-1}X_1(l)\sum_{n=0}^{N-1}x_2(n)W_N^{(k-l)n}$$

$$Y(k) = \frac{1}{N}\sum_{l=0}^{N-1}X_1(l)X_2(k-l). \qquad \square$$

(vi) Parseval's Relation: If the DFT of $x(n)$ is $X(k)$ the Parseval's relation is given by $\sum_{n=0}^{N-1}|x(n)|^2 = \frac{1}{N}\sum_{k=0}^{N-1}|X(k)|^2$.

Proof.

$$\sum_{n=0}^{N-1}|x(n)|^2 = \sum_{n=0}^{N-1}x(n)x^*(n) = \sum_{n=0}^{N-1}\frac{1}{N}\sum_{k=0}^{N-1}X(k)W_N^{-kn}x^*(n)$$

$$= \frac{1}{N}\sum_{k=0}^{N-1}X(k)\sum_{n=0}^{N-1}x^*(n)W_N^{-kn}$$

$$= \frac{1}{N}\sum_{k=0}^{N-1}X(k)\left(\sum_{n=0}^{N-1}x(n)W_N^{kn}\right)^*$$

$$= \frac{1}{N}\sum_{k=0}^{N-1}X(k)X^*(k) = \frac{1}{N}\sum_{k=0}^{N-1}|X(k)|^2. \qquad \square$$

3.4 Circular Convolution

The result of a circular convolution of two sequences each of length N is also of length N. This is why it is referred to as N-point circular convolution. To emphasize that it is N-point and to differentiate it with linear convolution the notation used for circular convolution of two sequences $x(n)$ and $h(n)$, each of length N is $x(n) \circledN h(n)$.

3.4.1   Graphical Implementation

In Section 3.3, we defined a circular shift to account for the shift of samples in a finite length sequence that has been defined over a specified length. Samples cannot be shifted into an undefined range and hence they are shifted in a cyclic manner. For a right shift the sample shifted from the last location move to the first location and every other sample moves one position in a counter-clockwise direction. Equation (3.9) represents circular convolution.

$$y(n) = \sum_{m=0}^{N-1} x(m)h(\langle n - m \rangle_N).$$
(3.9)

In the implementation of circular convolution, there is time reversal and circular shift of one sequence, in this case when n is zero, $h(\langle -m \rangle_N)$. For the time reversal of the sequence $h(n)$, the sample $h(0)$ retains its location as a reference and the remaining samples are mapped in the mirror image as $h(-n)$. From here the time reversed sequence is circularly shifted n samples to the right as $h(\langle n - m \rangle_N)$. For each shift n, the sum of products is obtained to give the output sample $y(n)$. This operation is clarified in Example 3.7.

Example 3.7. Determine the circular convolution of two length 4 sequences given by $w(n) = [2, 1.5, 1, 0.5]$ and $v(n) = [1, 0, 1, 0]$ for $0 \le n \le 3$

$$y(n) = \sum_{m=0}^{N-1} v(m)w(\langle n - m \rangle_N).$$

The two sequences represented graphically are shown in Figure 3.5.

3.4.2   Computation using Matrices

In this section, we will show the computation of circular convolution using matrices for two length-4 sequences. The expression for circular convolution of a sequence $h(n)$ and another sequence $x(n)$ each of

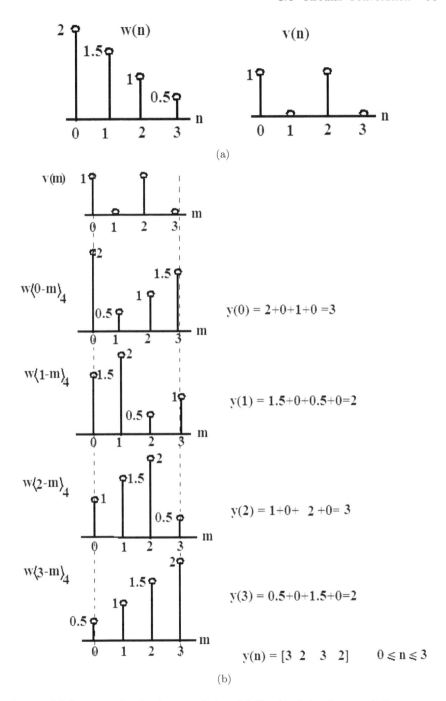

Fig. 3.5 (a) Sequences for circular convolution. (b) Graphical circular convolution.

length 4 is derived from Equation (3.9) and is given by

$$y(n) = \sum_{m=0}^{3} x(m)h(\langle n - m \rangle_4). \qquad (3.10)$$

This equation can be expressed in matrix form as follows:

$$\begin{bmatrix} y(0) \\ y(1) \\ y(2) \\ y(3) \end{bmatrix} = \begin{bmatrix} h(0) & h(\langle -1 \rangle_4) & h(\langle -2 \rangle_4) & h(\langle -3 \rangle_4) \\ h(1) & h(0) & h(\langle -1 \rangle_4) & h(\langle -2 \rangle_4) \\ h(2) & h(1) & h(0) & h(\langle -1 \rangle_4) \\ h(3) & h(2) & h(1) & h(0) \end{bmatrix} \begin{bmatrix} x(0) \\ x(1) \\ x(2) \\ x(3) \end{bmatrix}.$$

If we use the relation that $h(\langle -n \rangle_N) = h(N - n)$, we get

$$\begin{bmatrix} y(0) \\ y(1) \\ y(2) \\ y(3) \end{bmatrix} = \begin{bmatrix} h(0) & h(3) & h(2) & h(1) \\ h(1) & h(0) & h(3) & h(2) \\ h(2) & h(1) & h(0) & h(3) \\ h(3) & h(2) & h(1) & h(0) \end{bmatrix} \begin{bmatrix} x(0) \\ x(1) \\ x(2) \\ x(3) \end{bmatrix}. \qquad (3.11)$$

It is very easy to generate the elements of the H-matrix in Equation (3.11) if one observes that the elements in each diagonal are equal and that each succeeding row is a circular shift of previous row. We can show how to apply this by re-doing Example 3.7 in Example 3.8 using the concept of a circulant matrix.

Example 3.8. Determine the circular convolution of two length-4 sequences given by $x(n) = [2\ 1.5\ 1\ 0.5]$ and $h(n) = [1\ 0\ 1\ 0]$ for $0 \le n \le 3$ using a circulant matrix.

Solution

$y(n) = \sum_{m=0}^{3} x(m)h(\langle n - m \rangle_4)$ can be written in matrix form using the concept of a circulant matrix as

$$\begin{bmatrix} y(0) \\ y(1) \\ y(2) \\ y(3) \end{bmatrix} = \begin{bmatrix} 1 & 0 & 1 & 0 \\ 0 & 1 & 0 & 1 \\ 1 & 0 & 1 & 0 \\ 0 & 1 & 0 & 1 \end{bmatrix} \begin{bmatrix} 2 \\ 1.5 \\ 1 \\ 0.5 \end{bmatrix} = \begin{bmatrix} 3 \\ 2 \\ 3 \\ 2 \end{bmatrix}.$$

3.4.3   MATLAB Computation of Circular Convolution

The MATLAB function that is used to implement circular convolution is CCONV and the syntax is $c = \mathrm{cconv}(a, b, n)$, where a and b are the vectors to be convolved and n is the length of the resulting vector. If n is omitted the function defaults to length(a) + length(b) − 1 and the circular convolution becomes equivalent to linear convolution.

Program 3.4

```
% Program to compute circular convolution
x = input('Input first sequence as a row matrix x = ');
h = input('Input second sequence as a row matrix h = ');
N = input('Desired length of output sequence N = ');
k = 0:1:N-1;
% Plot the first Sequence
subplot(3,1,1)
stem(k,x,'o');
title('First sequence');
xlabel('Time index n'); ylabel('Amplitude');
% Plot the second Sequence
subplot(3,1,2)
stem(k,h,'o');
title('First sequence');
xlabel('Time index n'); ylabel('Amplitude');
c = cconv(x,h,N);
subplot(3,1,3)
stem(k,c,'o');
title('First sequence');
xlabel('Time index n'); ylabel('Amplitude');
```

Example 3.9. Determine the circular convolution of two length-4 sequences given by $x(n) = [2\ 1.5\ 1\ 0.5]$ and $h(n) = [1\ 0\ 1\ 0]$ for $0 \le n \le 3$ using MATLAB functions.

Solution

Use Program 3.4 above to obtain the following results (see Figure 3.6).

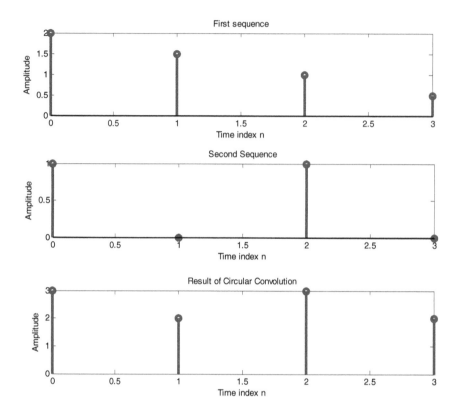

Fig. 3.6 MATLAB implementation of circular convolution.

3.4.4   DFT Implementation of Circular Convolution

The advantage of implementing circular convolution using the DFT route is due to the use of the efficient and fast FFT algorithm that is employed in the computation of the DFT. The DFTs of the time domain sequence are multiplied and their IDFT is computed to obtain the circular convolution. This is summarized in Figure 3.7.

Programs 3.1 and 3.2 can be modified and used to implement circular convolution.

3.5   The Fast Fourier Transform

The computation of the DTFT using DSP processors is not feasible as the DTFT is a continuous function of frequency that would require

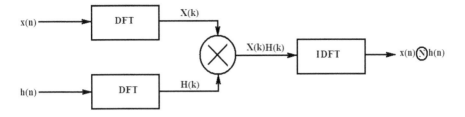

Fig. 3.7 DFT implementation of circular convolution.

an infinite amount of memory and infinite time to process an infinite number of samples. In order to solve this problem a finite number of samples of the DTFT in the frequency domain are taken for processing in what is referred to as the DFT.

The DFT is a finite length transform of a finite length sequence and would require a finite amount of memory to store the samples and a finite amount of time to process the samples. Sometimes the length of the sequence and the transform are too long such that the processing time required is too long and the memory required to store the samples is too large. Such sequences cannot be processed in DSP processors in real-time. It is possible to use some properties of the DFT to reduce the computation time and the memory required. The transforms that are formed after such simplification are referred to as the Fast Fourier Transforms (FFT). There are several version of the FFTs but we will focus only on the Decimation in Time FFT algorithm.

In order to develop the FFT algorithm, we will first look at the properties of the complex function W_N^{kn} that can be used for its simplification.

Properties the complex function W_N^{kn}

(i) If N is the length of the sequence then

$$W_N^{nN} = W_N^{kN} = e^{-j2\pi kN/N} = e^{-j2\pi k} = 1 = W_N^0. \qquad (3.12)$$

(ii) The complex function W_N^{kn} is periodic with respect to n.

$$W_N^{k(n+rN)} = W_N^{kn} W_N^{rkN} = W_N^{kn} \quad \text{for} \ -\infty < r < \infty. \qquad (3.13)$$

(iii) The complex function W_N^{kn} is periodic with respect to k.

$$W_N^{n(k+rN)} = W_N^{kn}W_N^{rnN} = W_N^{kn} \quad \text{for} \quad -\infty < r < \infty.$$
(3.14)

(iv) The even samples of the N-point complex sequence W_N^{kn} can be expressed as $N/2$-point sequence $W_{N/2}^{kr}$, where $n = 2r$.

$$W_N^{kn} = W_N^{2kr} = e^{2(-j2\pi kr/N)} = e^{-j2\pi kr/N/2} = W_{N/2}^{kr}. \quad (3.15)$$

Using these properties it is easy to show that a sequence of length N has a DFT that is periodic, i.e.,

$$X(k+N) = X(k). \quad (3.16)$$

3.5.1   The Decimation-in-Time FFT Algorithm

The Fast Fourier Transform (FFT) is used to compute an N-Point DFT by computing smaller-size DFTs and taking advantage of the periodicity and symmetry properties of the complex function W_N^{kn}. Consider a sequence $x(n)$ of length N, where N is a power of 2 (i.e., $N = 2^\mu$). In Equation (3.3) we have defined the DFT of a sequence $x(n)$ as

$$X(k) = \sum_{n=0}^{N-1} x(n)W_N^{kn} \quad \text{for } 0 \le k \le N-1, \quad (3.17)$$

where $W_N = e^{-j2\pi/N}$.

The computation of $X(k)$ requires N^2 complex multiplications and $N(N-1)$ complex additions.

The process of decimation-in-time

(i) Computation of $X(k)$

If N is a power of 2 it is possible to decimate $x(n)$ into $N/2$-point sequences such that one has samples which are even numbered and the other has samples that are odd numbered.

$$X(k) = \sum_{n \text{ even}} x(n)W_N^{kn} + \sum_{n \text{ odd}} x(n)W_N^{kn}$$

or

$$X(k) = \sum_{r=0}^{\frac{N}{2}-1} x(2r)W_N^{k2r} + \sum_{r=0}^{\frac{N}{2}-1} x(2r+1)W_N^{k(2r+1)}. \quad (3.18)$$

Applying the properties of the complex function W_N^{kn} we can write

$$X(k) = \sum_{r=0}^{\frac{N}{2}-1} x(2r) W_{N/2}^{kr} + W_N^k \sum_{r=0}^{\frac{N}{2}-1} x(2r+1) W_{N/2}^{kr}$$
(3.19)

or

$$X(k) = X_o(k) + W_N^k X_1(k), \qquad (3.20)$$

where $X_0(k)$ and $X_1(k)$ are $N/2$-point DFTs and $x_0(n) = x(2r)$ and $x_1(n) = x(2r+1)$. For $N = 8$ eight equations can be written relating N-point DFT $X(k)$ to $N/2$-point DFTs $X_0(k)$ and $X_1(k)$.

$$\left.\begin{aligned}
X(0) &= X_0(0) + W_N^0 X_1(0) \\
X(1) &= X_0(1) + W_N^1 X_1(1) \\
X(2) &= X_0(2) + W_N^2 X_1(2) \\
X(3) &= X_0(3) + W_N^3 X_1(3) \\
\\
X(4) &= X_0(0) + W_N^4 X_1(0) \\
X(5) &= X_0(1) + W_N^5 X_1(1) \\
X(6) &= X_0(2) + W_N^6 X_1(2) \\
X(7) &= X_0(3) + W_N^7 X_1(3)
\end{aligned}\right\} \qquad (3.21)$$

A flow graph representation of Equation (3.21) is shown in Figure 3.8.

The computation of the N-point DFT using the modified scheme requires two $N/2$-point DFTs which are combined with N complex multiplications and N complex additions. The total number of complex multiplications is $2\left(\frac{N}{2}\right)^2 + N = \frac{N^2}{2} + N$ and the total number of complex additions is $2\left(\frac{N}{2} - 1\right)\left(\frac{N}{2}\right) + N = \frac{N^2}{2}$. Compared to the original computation using Equation (3.7) the percentage reduction for both complex multiplication and additions is close to 50% for large values of N.

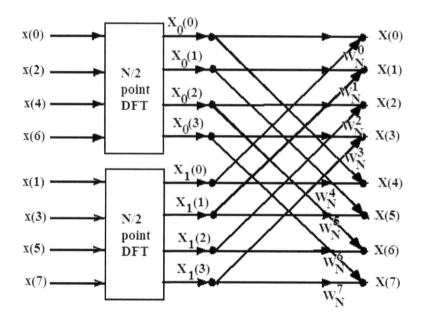

Fig. 3.8 Flow graph of first stage of decimation-in-time FFT algorithm.

(ii) Computation of $X_0(k)$

Further decimation of $x(n)$ can be achieved as long as $N/2$ is a power of 2. It is possible to express $X_0(k)$ and $X_1(k)$ in terms of $N/4$-point DFTs as follows:

$$X_0[k] = X_{00}[\langle k \rangle_{N/4}] + W_{N/2}^k X_{01}[\langle k \rangle_{N/4}], \qquad (3.22)$$

where $X_{00}(k)$ and $X_{01}(k)$ are $N/4$-point DFTs and $x_{00}(n) = x_0(2r) = x(4m)$ and $x_{01}(n) = x_0(2r) = x(4m+2)$. For $N = 8$ eight four equations can be written relating $N/2$-point DFT $X_0(k)$ to two $N/4$-point DFTs $X_{00}(k)$ and $X_{01}(k)$.

$$\left.\begin{array}{l} X_0(0) = X_{00}(0) + W_{N/2}^0 X_{01}(0) \\ X_0(1) = X_{00}(1) + W_{N/2}^1 X_{01}(1) \\ X_0(2) = X_{00}(0) + W_{N/2}^2 X_{01}(0) \\ X_0(3) = X_{00}(1) + W_{N/2}^4 X_{01}(1) \end{array}\right\} \qquad (3.23)$$

A flow graph representation of Equation (3.22) is shown in Figure 3.9.

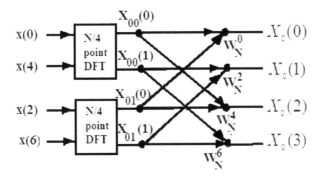

Fig. 3.9 Flow graph $N/2$-point decimation-in-time FFT algorithm to produce $X_0(k)$.

(iii) Computation of $X_1(k)$

$$X_1[k] = X_{10}[\langle k \rangle_{N/4}] + W_{N/2}^k X_{11}[\langle k \rangle_{N/4}], \qquad (3.24)$$

where $X_{10}(k)$ and $X_{11}(k)$ are $N/4$-point DFTs and $x_{10}(n) = x_1(2r) = x(4m+1)$ and $x_{11}(n) = x_1(2r+1) = x(4m+3)$. For $N = 8$ eight four equations can be written relating $N/2$-point DFT $X_1(k)$ to two $N/4$-point DFTs $X_{10}(k)$ and $X_{11}(k)$.

$$\left.\begin{aligned}
X_1(0) &= X_{10}(0) + W_{N/2}^0 X_{11}(0) \\
X_1(1) &= X_{10}(1) + W_{N/2}^1 X_{11}(1) \\
X_1(2) &= X_{10}(0) + W_{N/2}^2 X_{11}(0) \\
X_1(3) &= X_{10}(1) + W_{N/2}^4 X_{11}(1)
\end{aligned}\right\} \qquad (3.25)$$

A flow graph representation of Equation (3.24) is shown in Figure 3.10.

(iv) The Combined Flow Graph

When the flow graphs for the computation of $X_0(k)$ (Figure 3.9) and $X_1(k)$ (Figure 3.10) are substituted into the flow graph for the computation of $X(k)$ (Figure 3.8), we get the flow graph of Figure 3.11.

(v) A Two-point DFT

For $N = 8$ the $N/4$-point DFT is a two-point DFT given by $X_{00}(k) = \sum_{n=0}^{N-1} x_{00}(n) W_N^{kn} = \sum_{n=0}^{1} x_{00}(n) W_2^{kn}$ for $0 \leq k \leq 1$,

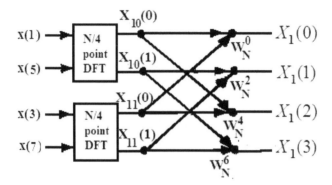

Fig. 3.10 Flow graph $N/2$-point decimation-in-time FFT algorithm to produce $X_1(k)$.

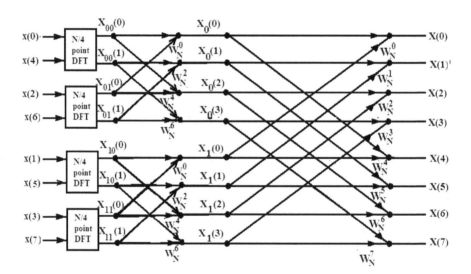

Fig. 3.11 Combined flow graph for the decimation-in-time FFT algorithm.

which can be simplified to

$$X_{00}(0) = x_{00}(0) + x_{00}(1) = x(0) + x(4), \quad \text{and}$$
$$X_{00}(1) = x_{00}(0) - x_{00}(1) = x(0) - x(4). \tag{3.26}$$

The flow graph to represent Equation (3.26) is shown in Figure 3.12.

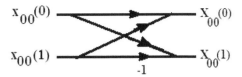

Fig. 3.12 A flow graph of a two-point DFT.

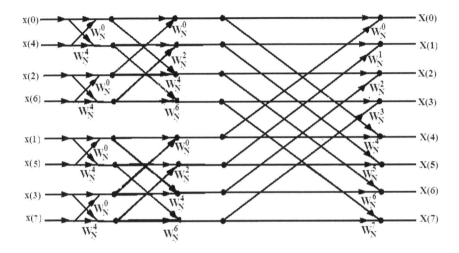

Fig. 3.13 The complete flow graph for the decimation-in-time FFT.

(vi) The Complete Flow Graph

The flow graph for the two point DFT can now substitute the $N/4$-point DFT in the flow graph of Figure 3.11 to give to form the flow graph shown in Figure 3.13.

It can be observed that the complete flow graph has three stages of computation; first to compute four two-point DFTs, then two four-point DFTs and finally one eight-point DFT. The number of computation stages depends on N and is given by $\log_2 N$. For $N = 8$, $\log_2 8 = 3$. On each computation stage there are N multiplications and N additions. Thus the original number of complex multiplications N^2 and the original number of complex additions $N(N - 1)$ is significantly reduced to $N \log_2 N$ for each operation. For large values of N

Table 3.2 Reduction in arithmetic operations.

N	Original number of additions and multiplications ($\approx N^2$)	Number of additions and multiplications after decimation in time ($= N \log_2 N$)	Reduction (%)
8	64	24	62.5
64	4096	384	90.6
512	262144	4608	98.2

the reduction in the number of complex operations is huge compared to when N is small as shown on Table 3.2.

(vii) Further Reduction

On careful observation of the structure of Figure 3.13 one notices that it is possible to reduce the number of multiplication further by doing one multiplication before branching rather than doing two multiplications in the branches. One must also use the fact that $W_8^7 = W_8^4 * W_8^3$, $W_8^6 = W_8^4 * W_8^2$, $W_N^5 = W_N^4 * W_N^1$, $W_N^4 = -1$. The new flow graph that is created is shown in Figure 3.14.

From Figure 3.14, we notice that the number has been reduced further by 50% to $\frac{N}{2} \log_2 N$. In fact when one takes into account $W_N^0 = 1$ the number of multiplications is reduced by more than 50%.

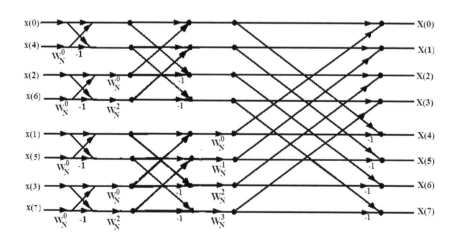

Fig. 3.14 The FFT flow graph with further reduction.

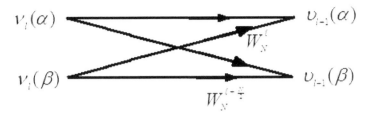

Fig. 3.15 The butterfly computational module.

3.5.2 Properties of the FFT Flow Graphs

(a) *The butterfly computation module*: It can be observed from Figure 3.13 that there is a basic two-input two-output computational unit that is repeated 8 times in each stage. It has the structure of a butterfly and is referred to as a butterfly computational unit. The advantage of the butterfly computational unit is that the same butterfly function or subroutine can be called at each stage of computation thus making the FFT program simple and short. A general flow graph for the butterfly computational unit can be developed if we make the following observations:

(i) There are N input variables and N output variables at each stage of computation.

(ii) If the number of computational stages is l then $0 \leq l \leq \mu$, where $\mu = \log_2 N$.

(iii) Define the input variable to the butterfly as $\nu_l(m)$ and the output variable as and $\nu_{l+1}(m)$, where $0 \leq m \leq N - 1$. Notice the following identities $\nu_1(0) = x(0)$, $\nu_1(1) = x(4)$, $\nu_4(0) = X(0)$, and $\nu_4(1) = X(4)$.

A general computational butterfly is then given in Figure 3.15.

It should be noted that α and β are selected values of m. The parameter ℓ can be calculated as $\ell = \kappa \gamma$, where κ is N divided by the number of DFT computation points of each stage (2-points for stage 1, 4 for stage 2, and 8 for stage 3) and γ is the number of the butterfly counting from 0. The parameter ℓ cannot exceed $N/2$.

We can express the butterfly output as follows:

$$v_{\ell+1}(\alpha) = v_l(\alpha) + W_N^\ell v_l(\beta)$$
$$v_{\ell+1}(\beta) = v_l(\alpha) + W_N^{\ell+\frac{N}{2}} v_l(\beta). \tag{3.27}$$

Equation (3.27) can be simplified to

$$v_{\ell+1}(\alpha) = v_l(\alpha) + W_N^\ell v_l(\beta)$$
$$v_{\ell+1}(\beta) = v_l(\alpha) - W_N^\ell v_l(\beta). \tag{3.28}$$

With this simplification Figure 3.15 can be redrawn such that the two complex multiplications in the branches are replaced by a single complex multiplication before branching as shown in Figure 3.16.

When this simplification of the butterfly computational module is applied to the flow graph of Figure 3.13 the flow graph of Figure 3.14 is obtained. This confirms the earlier observation and simplification that were made to achieve the flow graph of Figure 3.14.

(b) *In-place computation*: It is clear that the same butterfly module is used for computation at each stage. The input that is used for computation in a current stage is not required in the next stage. The current output can be stored in the same memory location to replace the input. Such computation that makes use of the same memory for the input and output is referred to as in-place computation. It achieves big saving in memory usage.

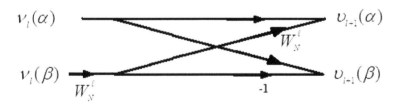

Fig. 3.16 A simplified butterfly computational module.

Table 3.3 Order of input samples.

Location of sample in decimal	Location of sample in binary	The sample number n in $x(n)$
0	000	000
1	001	100
2	010	010
3	011	110
4	100	001
5	101	101
6	110	011
7	111	111

(c) *Order of samples in input and output sequences:* It can be seen from Figure 3.14 that the order of the samples in the input sequence is not in sequential order as the output sequence. It may be difficult to predict where the samples go especially when the sequence length is large. In the examination of the order of the samples we can see the pattern reproduced in Table 3.3.

It easy to see that the sample that is located at $(n_1 n_2 n_3)$ is $x(n_3 n_2 n_1)$, where $n_1 n_2$ and n_3 are the bits that form the binary number representing the location. The location of the sample and the sample number are in bit reversed order.

(d) *MATLAB computation:* Programs 3.1 and 3.2 use the MATLAB fft and ifft functions to compute the DFT and the IDFT, respectively.

3.6 Problems

3.6.1 Suppose $x(n)$ is a complex sequence of length N with an N-point DFT $X(k)$. Find the DFTs of the following sequences in terms of $X(k)$.

(i) $x^*(n)$ (ii) $\text{Re}\{x(n)\}$ (iii) $j\text{Im}\,x(n)$.

3.6.2 Given a real $x(n)$ sequence of length N with an N-point DFT $X(k)$ prove the following symmetry relations

(i) $X(k) = X^*(\langle -k\rangle_N)$.

(ii) $\text{Re}\,X(k) = \text{Re}\,X(\langle -k\rangle_N)$.

(iii) $\text{Im}\,X(k) = -\text{Im}\,X(\langle -k\rangle_N)$.

3.6.3 Two length-4 sequences are given by $x(n) = [1,2,0,1]$ for $0 \le n \le 3$ and $h(n) = [2,1,0,1]$ for $0 \le n \le 3$. Determine the following

   (i) $X(k)$ and $H(k)$ for $0 \le k \le 3$,
   (ii) the product $Y(k) = X(k)H(k)$ for $0 \le k \le 3$,
   (iii) the sequence $y(n)$ the IDFT of $Y(k)$,
   (iv) the circular convolution of $x(n)$ and $h(n)$ using the circulant matrix.

3.6.4 A sequence $h(n)$ has a DFT $H(k) = \{1 + 2j, 0.5 - j, 3, 0.5 + j, 1 - 2j\}$. Obtain the DFT of the sequence $g(n)$ which is related to $h(n)$ according to $g(n) = h(\langle n - 2\rangle_5)$ using the DFT properties.

3.6.5 A length-8 sequence is given by $w(n) = \{2,0,1,3,2,1,4,1\}$. A DFT of another sequence $v(n)$ is related to the DFT of $w(n)$ according to $V(k) = W(\langle k - 4\rangle_8)$. Determine the sequence $v(n)$ without computing the DFTs.

3.6.6 A sequence is given by $w(n) = \{2,0,1,3,2,1,4,1\}$. Determine the following without computing the DFT

   (i) $X(0)$, (ii) $X(4)$, (iii) $\sum_{n=0}^{7} X(k)$, and (iv) $\sum_{n=0}^{7} |X(k)|^2$.

3.6.7 The 6 sample of a 12-point DFT of a real sequence $v(n)$ are given as $V(k) = \{1 + j, 3, 7 - j, 2, 3 + j, 1, ...\}$. Find the remaining DFT samples.

3.6.8 The DFT of a length-4 sequence is to be computed using the decimation-in-time FFT algorithm. Develop the flow graph for the computation from first principles. What is the minimum number of additions and multiplications that is achievable. Use the flow graph to compute the DFT of the sequence $x(n) = [1,0,1,0]$.

3.6.9 A program has been written to compute the DFT of a sequence. It is intended to use the same program to compute the inverse DFT of another sequence. Using block diagrams explain how this can be achieved.

3.6.10 Using MATLAB write a program to compute and plot the DFT magnitude and phase of any sequence of any length including sequences whose length are not power of 2.

4

The Transform Domain Analysis: The z-Transform

4.1 Introduction to the z-Transform

A powerful tool for the analysis of analog signals and systems is the Laplace transform. Through sampling it should be possible to develop a similar tool for discrete-time signals and systems.

The Laplace transform of an analog signal $x(t)$ is given by

$$X(s) = \int_{-\infty}^{\infty} x(t)e^{-st}dt, \qquad (4.1)$$

where $s = \sigma + j\Omega$.

If the analog signal is sampled at a rate $1/T$, where T is the sampling interval, to obtain a discrete-time signal then Equation (4.1) becomes

$$X'(s) = \sum_{n=-\infty}^{\infty} x'(nT)e^{-(\sigma+j\Omega)nT} = \sum_{n=-\infty}^{\infty} x'(nT)(e^{\sigma T}e^{j\Omega T})^{-n}. \qquad (4.2)$$

If we define $r = e^{\sigma T}, \theta = \Omega T$ radians (notice that θ has units of phase angle) and if we define the sample $x(n) = x'(nT)$ Equation (4.2) reduces to

$$X'(s) = \sum_{n=-\infty}^{\infty} x(n)(re^{j\theta})^{-n}.$$

We define the z-plane to be the plane spanned by $z = re^{j\theta}$ for $-\pi \leq \theta < \pi$ and $X'(s)$ becomes

$$X(z) = \sum_{m=-\infty}^{\infty} x(n)z^{-n}, \qquad (4.3)$$

where $X(z)$ represents the z-transform of the discrete-time signal $x(n)$.

Table 4.1 Some common z-transforms.

Sequence	z-transform	ROC
$\delta(n)$	1	All values of z
$\mu(n)$	$\dfrac{1}{1 - z^{-1}}$	$|z| > 1$
$\alpha^n \mu(n)$	$\dfrac{1}{1 - \alpha z^{-1}}$	$|z| > \alpha$
$-\alpha^n \mu(-n - 1)$	$\dfrac{1}{1 - \alpha z^{-1}}$	$|z| < \alpha$
$ce^{-\alpha n} \mu(n)$	$\dfrac{c}{1 - e^{-\alpha} z^{-1}}$	$|z| > e^{-\alpha}$
$r^n \cos(\omega_0 n) \mu(n)$	$\dfrac{1 - (r\cos(\omega_0))z^{-1}}{1 - (2r\cos(\omega_0))z^{-1} + r^2 z^{-2}}$	$|z| > r$
$r^n \sin(\omega_0 n) \mu(n)$	$\dfrac{1 - (r\sin(\omega_0))z^{-1}}{1 - (2r\cos(\omega_0))z^{-1} + r^2 z^{-2}}$	$|z| > r$

$X(z)$ is a complex variable in a complex plane. From Equation (4.3) we notice that the z-transform is represented by a power series. The z-transform will exist only where the power series converges. The region in the z-plane where the power series converges is called the region of convergence (ROC). We will look at z-transforms of some discrete-time signals and explore their regions of convergence (see Table 4.1).

Example 4.1. A unit sample sequence given by

$$x_1(n) = \delta(n) = \begin{cases} 1 & n = 0 \\ 0 & \text{elsewhere.} \end{cases} \tag{4.4}$$

The z-transform is given by

$$X_1(z) = \sum_{n=-\infty}^{\infty} x_1(n) z^{-n}$$

$$= \sum_{n=-\infty}^{\infty} \delta(n) z^{-n} = 1 \quad \text{for all values of } z. \tag{4.5}$$

ROC is everywhere on the z-plane.

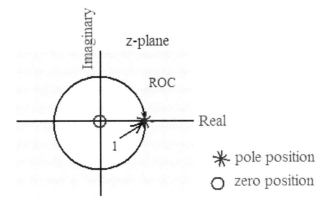

Fig. 4.1 ROC for a unit step sequence.

Example 4.2. A unit step sequence is given by

$$x_2(n) = \mu(n) = \begin{cases} 1 & n \geq 0 \\ 0 & n < 0. \end{cases} \tag{4.6}$$

The z-transform is given by

$$X_2(z) = \sum_{n=-\infty}^{\infty} x_2(n)z^{-n} = \sum_{n=-\infty}^{\infty} \mu(n)z^{-n}$$

$$= \sum_{n=0}^{\infty} z^{-n} = \frac{1}{1 - z^{-1}} \quad \text{for } z > 1. \tag{4.7}$$

ROC is the region for which $z > 1$. $X_2(z)$ has a zero at $z = 0$ and a pole at $z = 1$, Figure 4.1. ROC is bounded by the circle through the pole on the inside.

The unit step sequence is a causal sequence which means that unit step sequence has no nonzero valued samples for negative values on n. Notice that the ROC is exterior to a circle of unit radius.

Example 4.3. A causal exponential sequence is given by

$$x_3(n) = a^n \mu(n), \tag{4.8}$$

where $\mu(n)$ is a unit step sequence.

$$X_3(z) = \sum_{n=-\infty}^{\infty} x_3(n)z^{-n} = \sum_{n=-\infty}^{\infty} \alpha^n \mu(n)z^{-n} = \sum_{n=0}^{\infty} \alpha^n z^{-n}$$

$$= \sum_{n=0}^{\infty} (\alpha z^{-1})^n = \frac{1}{1 - \alpha z^{-1}} \quad \text{for } |\alpha z^{-1}| < 1. \tag{4.9}$$

Hence $X_3(z) = \frac{1}{1-\alpha z^{-1}}$ and ROC is $|z| > \alpha$.

$X_3(z)$ has a pole at $z = \alpha$ and a zero at $z = 0$. The ROC is similar to that of Figure 4.1 except that the pole position is at $z = \alpha$ instead of $z = 1$.

The causal exponential sequence has a region of convergence that is also exterior to a circle of radius α. In general, whenever a sequence is causal or right-sided, the region of convergence of its z-transform is always exterior to a circle of some specified radius.

A right-sided sequence is that sequence that has no nonzero valued samples to the left of a reference sample m. The reference sample m may be negative or positive. If it is positive then the sequence is causal.

Example 4.4. An anti-causal sequence is given by

$$x_4(n) = -\alpha^n \mu(-n - 1), \tag{4.10}$$

where $\mu(n)$ is a unit step sequence. Notice that $\mu(-n - 1)$ has unit values to the left of $n=0$ and has zero values elsewhere. The negative sign on the sequence has been intentionally placed there for the sake of comparison which will become clear after finding the z-transform.

$$X_4(z) = \sum_{n=-\infty}^{\infty} x_3(n)z^{-n} = \sum_{n=-\infty}^{\infty} -\alpha^n \mu(-n - 1)z^{-n}$$

$$= \sum_{n=-\infty}^{-1} -\alpha^n z^{-n}.$$

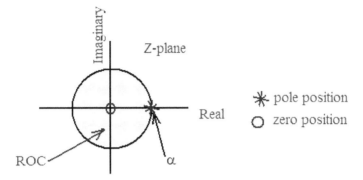

Fig. 4.2 ROC for an anti-causal exponential sequence.

Let $m = -n$

$$X_4(z) = \sum_{m=1}^{\infty} -\alpha^{-m} z^m = \alpha^{-1} z^1 \sum_{m=0}^{\infty} -\alpha^{-m} z^m$$

$$= -\alpha^{-1} z^1 \times \frac{1}{1 - \alpha^{-1} z^1} \quad \text{for } |\alpha^{-1} z^1| < 1,$$

which can be written as

$$X_4(z) = \frac{1}{1 - \alpha z^{-1}} \quad \text{for } |Z| < \alpha. \tag{4.11}$$

$X_4(z)$ has a zero at the origin and a pole at the $z = \alpha$. The ROC is bounded by the circle through the pole on the outside (see Figure 4.2).

Notice that the anti-causal sequence has a region of convergence that is interior to a circle of radius α. Whenever a sequence is anti-causal or left-sided the region of convergence of its z-transform is always interior to a circle of some specified radius.

A left-sided sequence is that sequence that has no nonzero valued samples to the right of a reference sample m. The reference sample m may be negative or positive. If it is negative then the sequence is anti-causal.

It should also be noted that the z-transforms of $X_3(z)$ and $X_4(z)$ are the same except for their regions of convergence. In order to specify the z-transform of a unique sequence the ROC must be given.

Example 4.5. Consider a two sided sequence given by

$$x_5(n) = \alpha^n \mu(n) - \beta^n \mu(-1-n).$$

From Examples 4.3 and 4.4 we note that the first term after the equality sign is similar to $x_3(z)$ and the second term is similar to $x_4(n)$. Therefore the z-transfrom of $x_5(n)$ is given by

$$X(z) = \frac{1}{1 - \alpha z^{-1}} + \frac{1}{1 - \beta z^{-1}} \quad \text{for } |z| > \alpha \quad \text{and for } |z| < \beta.$$

$$(4.12)$$

The region of convergence will be where the two regions of convergence overlap. Where there is no overlap then there is no region of convergence. It is also noted that ROC is bounded by the pole positions.

4.2   The Inverse z-Transform

The z-transform of the sequence $h(n)$ is given by

$$H(z) = \sum_{n=-\infty}^{\infty} h(n) z^{-n}. \qquad (4.13)$$

However, if we define $z = re^{j\omega}$ then we obtain

$$H(re^{j\omega}) = \sum_{n=-\infty}^{\aleph} h(n) r^{-n} e^{-j\omega n}.$$

This implies that the z-transform of $h(n)$ is the same as the Fourier transform of the modified sequence $h(n)r^{-n}$. We can obtain the inverse Fourier transform using previous results thus

$$h(n) r^{-n} = \frac{1}{2\pi} \int_{-\infty}^{\infty} H(re^{j\omega}) e^{j\omega n} d\omega.$$

If we substitute for $z = re^{j\omega}$ the above equation becomes a contour integral given by

$$h(n) = \frac{1}{2\pi j} \oint_{C'} H(z) z^{n-1} dz, \qquad (4.14)$$

where C' is a contour of integration in the counter-clockwise direction defined by $|z| = r$. The integral remains unchanged when C' is replaced by another C that encircles the point $z = 0$ in the region of convergence.

4.2.1   The Method of Residues [3]

The contour integral of Equation (4.14) can be evaluated using the Cauchy residue theorem given by

$$h(n) = \sum[\text{residue of } H(z)z^{n-1}$$

$$\text{at the pole of } H(z)z^{n-1} \text{ inside } C]. \qquad (4.15)$$

The residue of $H(z)z^{n-1}$ at the poles $z = p_l$ can be calculated from the formula

$$R_{z=p_l} = \frac{d^{m-1}}{dz^{m-1}} \left[\frac{(z - p_l)^m}{(m-1)!} H(z)z^{n-1} \right]_{z=p_l}, \quad m \geq 1, \qquad (4.16)$$

where m is the order of the pole at $z = p_l$.

This method of evaluating the inverse z-transform is referred to as the method of residues.

Example 4.6.   Determine the inverse z-transform of $X(z) = \frac{1}{(z-2)(z-4)}$ for $|z| > 2$ using the method of residues.

Solution

$$X(z)z^{n-1} = \frac{z^{n-1}}{(z-2)(z-4)}.$$

The function has a simple pole at $z = 0$ when $n = 0$ and no poles at $z = 0$ for $n > 0$. There are also simple poles at $z = 2$, and at $z = 4$ for all values of n. The residue at $z = 0$ is given by

$$\text{Res}_{z=0} = \frac{z}{1!} \frac{z^{n-1}}{(z-2)(z-4)} = \frac{1}{8} \quad \text{for } n = 0,$$

$$\text{Res}_{z=2} = \frac{z-2}{1!} \frac{z^{n-1}}{(z-2)(z-4)} = -\frac{1}{4} \quad \text{for } n = 0$$

and

$$\text{Res}_{z=4} = \frac{z-4}{1!} \frac{z^{n-1}}{(z-2)(z-4)} = \frac{1}{8} \quad \text{for } n = 0$$

$$x(0) = \frac{1}{8} - \frac{1}{4} + \frac{1}{8} = 0$$

For $n > 0$

$$\text{Res}_{z=2} = \frac{z-2}{1!} \frac{z^{n-1}}{(z-2)(z-4)} = -\frac{1}{2} 2^{n-1},$$

$$\text{Res}_{z=4} = \frac{z-4}{1!} \frac{z^{n-1}}{(z-2)(z-4)} = \frac{1}{2} 4^{n-1}.$$

$$x(n) = \text{Res}_{z=2} + \text{Res}_{z=4} = \frac{1}{2}(4^{n-1} - 2^{n-1}) \quad \text{for } n \geq 0 \quad \text{or}$$

$$= \frac{1}{2}(4^{n-1} - 2^{n-1})\mu(n).$$

4.2.2   Method using Partial Fraction Expansion [5]

There are other simpler methods for finding the inverse z-transform involving partial fraction expansion and long division. If $H(z)$ is rational function which is a z-transform of a causal sequence $h(n)$ then it may be easy to express $H(z)$ as a sum of partial fractions involving simpler terms whose inverse z-transforms can be read off from tables. Expressing $H(z)$ as rational function

$$H(z) = \frac{C(z)}{D(z)}, \tag{4.17}$$

where $C(z)$ and $D(z)$ are polynomials in z^{-1} of degree M and N, respectively. If the degree of $C(z)$ is greater than the degree of $D(z)$ then divide $C(z)$ by $D(z)$ and obtain

$$H(z) = \sum_{l=0}^{M-N} \gamma_l z^{-l} + \frac{C_1(z)}{D(z)}. \tag{4.18}$$

The ratio $C_1(z)/D(z)$, where the degree of numerator polynomial is less than the degree of the denominator polynomial is referred to as a proper fraction.

4.2.2.1   Partial Fraction Expansion of $H(z)$ with Simple Poles

We will first consider the case where $H(z)$ is a proper fraction with simple poles. Simple poles imply that there is only one distinct pole at each location.

Let N distinct poles of $H(z)$ be located at ξ_k for $1 \le k \le N$. The partial fraction expansion of H(z) is then given by

$$H(z) = \sum_{l=1}^{N} \frac{\lambda_l}{1 - \xi_l z^{-1}}, \tag{4.19}$$

where the constant λ_l, referred to as the residue is given by

$$\lambda_l = (1 - \xi_l) H(z)|_{z=\xi_l}. \tag{4.20}$$

If the ROC is exterior to a circle passing through ξ_l (i.e., $z > |\xi_l|$) then the inverse z-transform of $\frac{\lambda_l}{1-\xi_l z^{-1}}$ will be $\lambda_l(\xi_l)^n \mu(n)$. The inverse z-transform of $H(z)$ is finally given by

$$h(n) = \sum_{l=1}^{N} \lambda_l(\xi_l)^n \mu(n). \tag{4.21}$$

Notice that it is possible to obtain partial fraction expansions as a function of z (instead of z^{-1}). The approach leads to correct results but you will not be able to use the table of standard z-transforms provided in this book.

In the case where ROC is interior to a circle passing through ξ_l (i.e., $z < |\xi_l|$) then the inverse z-transform of $\frac{\lambda_l}{1-\xi_l z^{-1}}$ will be $-\lambda_l(\xi_l)^n \mu(-n - 1)$, see Example 4.4 in Section 4.1. The inverse z-transform of $H(z)$ is finally given by

$$h(n) = \sum_{l=1}^{N} -\lambda_l(\xi_l)^n \mu(-n - 1). \tag{4.22}$$

It may be possible that ROC is bounded by two poles (i.e., $\xi_k < |z| < \xi_j$, then the inverse z-transform will have a combination of causal and anti-causal sequences depending on whether ROC is exterior or interior to a circle through the specific pole (see Figure 4.3).

Example 4.7. Determine the inverse z-transform of $X(z) = \frac{1}{(z-2)(z-4)}$ for $|z| > 2$ using the method of partial fraction expansion.

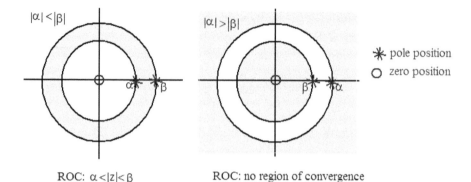

ROC: $\alpha < |z| < \beta$ ROC: no region of convergence

Fig. 4.3 ROC for a sum of a causal and an anti-causal sequences.

Solution

Expressing $X(z)$ as a function of z^{-1} we get

$$X(z) = \frac{z^{-2}}{(1 - 2z^{-1})(1 - 4z^{-1})} \quad \text{for } |z| > 2.$$

Both the numerator and the denominator are second-order polynomials of z^{-1} and $X(z)$ must be converted to a proper fraction by long division giving

$$X(z) = \frac{1}{8}\left(1 + \frac{6z^{-1} - 1}{(1 - 2z^{-1})(1 - 4z^{-1})}\right).$$

The second term can be expanded using partial fraction expansion as

$$\left(\frac{6z^{-1} - 1}{(1 - 2z^{-1})(1 - 4z^{-1})}\right) = \frac{\lambda_1}{1 - 2z^{-1}} + \frac{\lambda_2}{1 - 4z^{-1}}.$$

$$\lambda_1 = (1 - 2z^{-1})\left(\frac{6z^{-1} - 1}{(1 - 2z^{-1})(1 - 4z^{-1})}\right)\Big|_{z^{-1} = \frac{1}{2}} = -2.$$

$$\lambda_2 = (1 - 4z^{-1})\left(\frac{6z^{-1} - 1}{(1 - 2z^{-1})(1 - 4z^{-1})}\right)\Big|_{z^{-1} = \frac{1}{4}} = 1.$$

$$X(z) = \frac{1}{8}\left(1 + \frac{-2}{(1 - 2z^{-1})} + \frac{1}{(1 - 4z^{-1})}\right).$$

$$x(n) = \frac{1}{8}\delta(n) - \frac{1}{4}(2)^n\mu(n) + \frac{1}{8}(4)^n\mu(n).$$

This result can be shown to be equal to the result of Example 4.6. Notice that $x(0) = 0$ giving the same answer as with the residual method. For $n > 0$, $\delta(n) = 0$ and the two expressions for $x(n)$ obtained by the two methods are identical.

4.2.2.2 Partial Fraction Expansion with Multiple Poles

If in general the rational fraction $H(z) = C(z)/D(z)$ is an improper fraction with the polynomial $C(z)$ having order M it can be reduced to an expression $H(z) = \sum_{l=0}^{M-N-L} \gamma_l z^{-l} + \frac{C_1(z)}{D(z)}$, using long division, in such a way that the ratio $C_1(z)/D(z)$, is a proper fraction with the polynomial $C_1(z)$ having order $N - L$. It is possible that $D(z)$ may contain identical multiple poles at $z = \xi_v$. Let the number of identical poles at $z = \xi_v$ be equal to L and the remaining poles be simple poles. We can express $H(z)$ as follows:

$$H(z) = \sum_{l=0}^{M-N-L} \gamma_l z^{-l} + \sum_{l=1}^{N-L} \frac{\lambda_l}{1 - \xi_l z^{-1}} + \sum_{i=1}^{L} \frac{\chi_i}{(1 - \xi_v z^{-1})^i}, \quad (4.23)$$

where the constants λ_i are the residues and are computed in the same manner as for simple poles, the constants χ_I are computed using the formula

$$\chi_i = \frac{1}{(L - i)!(-\xi_v)^{L-1}} \frac{d^{L-1}}{d(z^{-1})^{L-1}} \times \left[(1 - \xi_v z^{-1})^L H(z) \right]|_{z=\xi_v} \quad \text{for } 1 \leq i \leq L. \quad (4.24)$$

4.2.2.3 Method using Long Division

The z-transform of a causal sequence $h(n)$ is a power series of z^{-1} as seen in (4.25)

$$H(z) = \sum_{n=0}^{\infty} h(n)z^{-n}. \quad (4.25)$$

The nth sample of the sequence $h(n)$ is the coefficient of z^{-n}. Hence if a rational function $H(z) = C(z)/D(z)$ is expressed as a power series of z^{-n} the coefficients of the series represent the sequence. $H(z)$ can be expressed as a power series by long division.

Example 4.8. Determine the inverse z-transform of $X(z) = \frac{1}{(z-2)(z-4)}$ for $|Z| > 2$ using the method of long division.

Solution

We can write

$$X(z) = \frac{1}{(z-2)(z-4)} = \frac{z^{-2}}{1 - 6z^{-1} + 8z^{-2}}$$

$$
\begin{array}{r}
z^{-2} - 6z^{-3} + 28z^{-4} + 120z^{-5} + \\
\hline
1 - 6z^{-1} + 8z^{-2} \overline{\smash{\big)}\, z^{-2}} \\
\end{array}
$$

$$
\begin{array}{r}
z^{-2} - 6z^{-3} + 8z^{-4} \\
+6z^{-3} - 8z^{-4} \\
6z^{-3} - 36z^{-4} + 48z^{-5} \\
+ 28z^{-4} - 48z^{-5} \\
+ 28z^{-4} - 168z^{-5} + 224z^{-6} \\
+ 120z^{-5} - 224z^{-6} \\
+ 120z^{-5} - 720z^{-6} + 960z^{-7} \\
496z^{-6} - 960z^{-7} \\
\end{array}
$$

The inverse z-transform of $X(z)$ is then given by

$$x(n) = \delta(n-2) + 6\delta(n-3) + 28\delta(n-4) + 120\delta(n-5) + \cdots.$$

The long division method does not give a closed form expression while the methods of residue and partial fraction expansion give a closed form expression.

4.3   Properties of z-transforms

In this section, we give the properties of z-transforms and show the proofs of a selected number of them. The proofs of the rest are left as an exercise to the student. The properties simplify the implementations of certain concepts in the design and implementation of discrete-time systems (see Table 4.2).

(i) The Linearity Property: The z-transform is a linear transformation. This can be proved by showing that the z-transform

Table 4.2 A summary of useful properties of the z-transform [3].

	Sequence	z-transform	ROC
Linearity	$\alpha x(n) + \beta y(n)$	$\alpha X(z) + \beta X(z)$	$R_x \cap R_y$
Time reversal	$x(-n)$	$X(1/z)$	$1/R_x$
Time-shifting	$x(n - n_0)$	$z^{-n_0} X(z)$	R_x except $z = 0$
Multiplication by an exponential sequence	$a^n x(n)$	$X(z/a)$	$|a| R_x$
Differentiation	$n x(n)$	$-z \dfrac{dX(z)}{dz}$	R_x except $z = \infty$
Convolution	$x(n) \otimes y(n)$	$X(z)Y(z)$	$R_x \cap R_y$
Conjugation	$x * (n)$	$X^*(z^*)$	R_x

Let $X(z) = Z\{x(n)\}$ with ROC R_x, and $Y(z) = Z\{y(n)\}$ with ROC R_y.

of a linear combination of two sequences is a linear combination of their z-transforms.

Proof. Let the linear combination of two sequences $x_1(n)$ and $x_2(n)$ be given by $y(n) = \alpha x_1(n) + \beta x_2(n)$. Let the z-transform of $x_1(n)$ and $x_2(n)$ be $X_1(z) = \sum_{n=-\infty}^{\infty} x_1(n) z^{-n}$ and $X_2(z) = \sum_{n=-\infty}^{\infty} x_2(n) z^{-n}$. The z-transform of $y(n)$ is given by

$$Y(z) = \sum_{n=-\infty}^{\infty} y(n) z^{-n} = \sum_{n=-\infty}^{\infty} (\alpha x_1(n) + \beta x_2(n)) z^{-n}$$

$$= \alpha \sum_{n=-\infty}^{\infty} x_1(n) z^{-n} + \beta \sum_{n=-\infty}^{\infty} x_2(n) z^{-n}$$

$\alpha X_1(z) + \beta X_2(z)$ *which is a linear combination of the z-transforms of* $x_1(n)$ *and* $x_2(n)$. ☐

(ii) The Time Shifting Property: If a sequence $x(n)$ with a z-transform $X(z)$ is shifted in time by n_0, to obtain a new sequence $x_1(n) = x(n - n_0)$, then the z-transform of the new sequence is given by $X_1(z) = z^{-n_0} X(z)$.

Proof. By definition

$$X(z) = \sum_{n=-\infty}^{\infty} x(n) z^{-n}$$

and also

$$X_1(z) = \sum_{n=-\infty}^{\infty} x_1(n)z^{-n} = \sum_{n=-\infty}^{\infty} x(n-n_0)z^{-n}.$$

Let $m = n - n_0$. Then

$$X_1(z) = \sum_{n=-\infty}^{\infty} x(m)z^{-(m+n_0)}$$

$$= z^{-n_0} \sum_{n=-\infty}^{\infty} x(m)z^{-m} = z^{-n_0} X(z). \qquad \square$$

(iii) Differentiation Property: If a sequence $x(n)$ has a z-transform given by $X(z)$ then the sequence $nx(n)$ has a z-transform given by $-z\frac{dX(z)}{dz}$.

Proof. By definition $X(z) = \sum_{n=-\infty}^{\infty} x(n)z^{-n}$. Differentiating with respect to z gives $\frac{dX(z)}{dz} = \sum_{n=-\infty}^{\infty} -nx(n)z^{-n-1}$, which can be written as $-z\frac{dX(z)}{dz} = \sum_{n=-\infty}^{\infty} nx(n)z^{-n}$. This shows that $-z\frac{dX(z)}{dz}$ is the z-transform of $nx(n)$. $\qquad \square$

(iv) The Convolution Property: If a sequence $x_1(n)$ has a z-transform $X_1(z)$ and a sequence $x_2(n)$ has a z-transform $X_2(z)$ then the z-transform of the sequence $x_1(n) \otimes x_2(n)$, (i.e., convolution of $x_1(n)$ and $x_2(n)$) is given by $X_1(z)X_2(z)$.

Proof. Let

$$X_1(z) = \sum_{n=-\infty}^{\infty} x_1(n)z^{-n}$$

and

$$X_2(z) = \sum_{n=-\infty}^{\infty} x_2(n)z^{-n}$$

$$X_3(z) = \sum_{n=-\infty}^{\infty} x_1(n) \otimes x_2(n)z^{-n}$$

$$= \sum_{n=-\infty}^{\infty} \sum_{k=-\infty}^{\infty} x_1(n-k)x_2(k)z^{-n}.$$

Changing the order of summation we get

$$X_3(z) = \sum_{k=-\infty}^{\infty} \sum_{n=-\infty}^{\infty} x_1(n-k)x_2(k)z^{-n}$$

$$= \sum_{k=-\infty}^{\infty} x_2(k) \sum_{n=-\infty}^{\infty} x_1(n-k)z^{-n}.$$

Substituting $m = n - k$ we get

$$X_3(z) = \sum_{k=-\infty}^{\infty} x_2(k) \sum_{n=-\infty}^{\infty} x_1(m)z^{-(m+k)}$$

$$= \sum_{k=-\infty}^{\infty} x_2(k)z^{-k} \sum_{n=-\infty}^{\infty} x_1(m)z^{-m} = X_2(z)X_1(z) \qquad \square$$

(v) **Multiplication by an Exponential Sequence:** If a sequence $x(n)$ has a z-transform $X(z)$, then a sequence $\alpha^n x(n)$ has a z-transform $X(z/\alpha)$.

Proof. Let

$$X(z) = \sum_{n=-\infty}^{\infty} x(n)z^{-n}$$

and

$$X_1(z) = \sum_{n=-\infty}^{\infty} \alpha^n x(n)z^{-n}$$

$$= \sum_{n=-\infty}^{\infty} x(n) \left(\frac{z}{\alpha}\right)^{-n} = X\left(\frac{z}{\alpha}\right) \qquad \square$$

(vi) **Time Reversal Property:** If a sequence $x(n)$ has a z-transform $X(z)$, then a sequence $x(-n)$ has a z-transform $X(1/z)$

Proof. If a sequence $x(n)$ has a z-transform $X(z)$, then a sequence $x(-n)$ has a z-transform

$$X_1(z) = \sum_{n=-\infty}^{\infty} x(-n)z^{-n}.$$

Let $m = -n$. Then

$$X_1(z) = \sum_{n=-\infty}^{\infty} x(m)z^m = \sum_{n=-\infty}^{\infty} x(m)\left(\frac{1}{z}\right)^{-m} = X\left(\frac{1}{z}\right). \quad \square$$

(vii) The Conjugation Property: If a sequence $x(n)$ has a z-transform $X(z)$, then a sequence $x^*(n)$ has a z-transform $X^*(z^*)$.

Proof. If a sequence $x(n)$ has a z-transform $X(z)$, then a sequence $x^*(n)$ has a z-transform

$$X_1(z) = \sum_{n=-\infty}^{\infty} x^*(n)z^{-n}$$

$$= \left(\sum_{n=-\infty}^{\infty} x(n)(z^*)^{-n}\right)^* = X^*(z^*). \quad \square$$

4.4   Transfer Functions of Discrete-Time Systems

An LTI discrete-time system is characterized by a linear constant coefficients difference equation given by

$$\sum_{k=0}^{N} a_k y(n-k) = \sum_{k=0}^{M} b_k x(n-k), \qquad (4.26)$$

where $x(n)$ is the input and $y(n)$ is the output of the system, a_k and b_k are constant coefficients. The order of the system is the $\max(N, M)$. You can solve for the current output by making $y(n)$ the subject of the formula.

$$y(n) = \sum_{k=0}^{M} b_k x(n-k) - \sum_{k=1}^{N} a_k y(n-k). \qquad (4.27)$$

Such a system has an impulse response of infinite length (IIR) but is realizable since it is implemented using a finite sum of products terms from the linear constant coefficient difference equation. It would not be realizable if it is implemented using the convolution sum of the input and the infinite length impulse response.

We can obtain the z-transform of the system by using the linearity and the time-shifting properties as follows:

$$Y(z)\sum_{k=0}^{N}a_k z^{-k} = X(z)\sum_{k=0}^{M}b_k z^{-k}. \tag{4.28}$$

The transfer function of the system is obtained as

$$H(z) = \frac{Y(z)}{X(z)} = \frac{\sum_{k=0}^{M}b_k z^{-k}}{\sum_{k=0}^{N}a_k z^{-k}}$$

$$= \frac{b_0 + b_1 z^{-1} + b_2 z^{-2} + \cdots + b_M z^{-M}}{a_0 + a_1 z^{-1} + a_2 z^{-2} + \cdots + a_N z^{-N}}. \tag{4.29}$$

If all the denominator coefficients were zero except $a_0 = 1$, $H(z)$ will have a transfer function given by $H(z) = b_0 + b_1 z^{-1} + b_2 z^{-2} + \cdots + b_M z^{-M}$. This represents the transfer function of a Finite Impulse Response (FIR) filter. In this case the impulse response will be of finite length and the coefficients of z^{-1} represent the impulse response samples. Instead of using the numerator coefficient of $H(z)$ we will use the values of the impulse response to write the z-transform of the FIR system.

$$H(z) = h(0) + h(1)z^{-1} + h(2)z^{-2} + h(3)z^{-3}\ldots h(M-1)z^{M-1}. \tag{4.30}$$

In the next section, we show how Equations (4.29) and (4.30) can be used to obtain the realization diagrams for the IIR filters and FIR filters, respectively. This is made possible because with the z-transform it is possible to make simple algebraic manipulations.

4.5 Poles and Zeros

The transfer function of a discrete-time system $H(z)$ given by Equation (4.29) may also be written in factored form as

$$H(z) = \frac{Y(z)}{X(z)} = \frac{\sum_{k=0}^{M}b_k z^{-k}}{\sum_{k=0}^{N}a_k z^{-k}} = \beta\frac{\prod_{l=1}^{M}(1-\xi_l z^{-1})}{\prod_{l=1}^{N}(1-\gamma_l z^{-1})}. \tag{4.31}$$

The roots of the numerator polynomial are called zeros since $H(z)|_{z=\xi_l} = 0$. $H(z)$ has M zeros at $z = \xi_l$ for $1 \le l \le M$. The roots of the denominator polynomial are called poles since $H(z)|_{z=\xi_l} = \infty$. $H(z)$ has N roots at $z = \gamma_l$ for $1 \leq l \leq N$. The LTI system described in Equation (4.31) is a pole–zero system while the system described by Equation (4.30) is an all zero system. The poles and zeros may be real or complex. When they are complex they appear in conjugate pairs. This is because the coefficients of the transfer function must be real for the system to be realizable. Sometimes there may be multiple poles.

A MATLAB function z-plane (b, a) can be used to plot the poles and zeros on the z-plane and relative to a unit circle. The parameters b and a represent the row matrices of the numerator and denominator coefficients of the transfer function, respectively. The location of the poles relative to unit circle gives information about the stability of the system. For a system to be stable all poles must lie inside the unit circle.

Example 4.9. A causal IIR transfer function is given by $H(z) = \frac{3z^3+2z^2+5}{(0.5z+1)(z^2+z+0.6)}$. Determine the stability condition of $H(z)$.

Solution

Find the poles with respect to the unit circle. Factorize the denominator. Denominator polynomial $= (0.5z + 1)(z + 0.5 + j0.59)(z + 0.5 - j0.59)$. Pole positions at $z = -2$, $z = -0.5 - j0.59$ and $-0.5 + j0.59$ Pole–zero plot showing only poles.

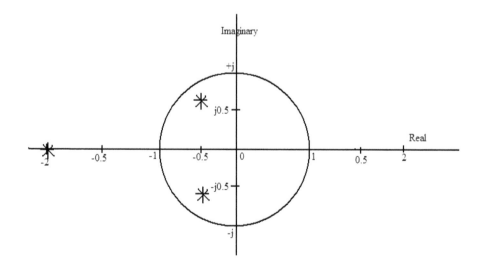

There is a pole outside the unit circle and therefore the system is unstable.

4.6   Realization Structures

In Section 1.3.2, we were introduced to the basic building blocks for discrete-time systems that were used to implement arithmetic operations and basic processes like delay. It is possible to use the same building blocks to realize different structures or block diagram representation from the transfer functions such as those represented by Equations (4.29) and (4.30). There are several advantages of using block diagrams and these include:

   (i) It is easy to write down the computational algorithm from the block diagram by inspection.
  (ii) It is easy to develop the input–output relation.
 (iii) It is easy to manipulate the block diagram to derive other equivalent structures.
 (iv) It is easy to determine the hardware requirements from the block diagram.

It is possible to realize many different structures from the same transfer function. Such structures will have identical performance if the filter implementation is done with infinite precision. However, the coefficients, the input and output signal and the intermediate signal variables are quantized or we can say that the signal processing operations within the DSP are done with finite precision. The different realization structures behave differently under finite precision arithmetic and it is up to the designer to determine the best structure. It is easier to do this through simulation.

4.6.1   Finite Impulse Response (FIR) filter

The FIR filter is represented by Equation (4.30). If we select a filter order of 4 then the transfer function becomes

$$H(z) = h(0) + h(1)z^{-1} + h(2)z^{-2} + h(3)z^{-3} + h(4)z^{-4}. \qquad (4.32)$$

If we take the inverse z-transform of Equation (4.32) we get the impulse response of the filter which is given by

$$h(n) = h(0)\delta(n) + h(1)\delta(n-1) + h(2)\delta(n-2)$$
$$+ h(3)\delta(n-3) + h(4)\delta(n-4). \tag{4.33}$$

To obtain the output we must convolve the input and the impulse response to get

$$y(n) = h(n) \otimes x(n)$$
$$= (h(0)\delta(n) + h(1)\delta(n-1) + h(2)\delta(n-2)$$
$$+ h(3)\delta(n-3) + h(4)\delta(n-4)) \otimes x(n)$$
$$= h(0)x(n) + h(1)x(n-1) + h(2)x(n-2)$$
$$+ h(3)x(n-3) + h(4)x(n-4). \tag{4.34}$$

The output $y(n)$ can be obtained by the structure of Figure 4.4.

The structure of Figure 4.4 is a direct form realization structure because the coefficients of the transfer function are the same as the multiplier coefficients in the realization structure. This structure is also known as the *tapped delay* line or a *transversal filter*. The number of delays is 4 and the order of the transfer function is 4. When the number of delays is the same as the order of the transfer function the structure is referred to as *canonic*.

It is possible to realize an equivalent structure by taking the transpose of this structure. The transpose operations is achieved through the following steps:

(i) Replace all pickoff points with summers and vice versa.
(ii) Change the directions of all arrows.

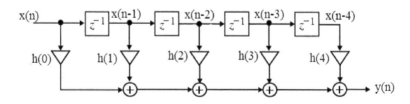

Fig. 4.4 Direct form FIR realization.

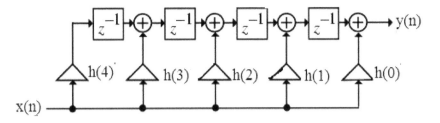

Fig. 4.5 The transpose of the direct form FIR realization.

(iii) Interchange input and output.
(iv) Rotate the figure to have the input on the left.

The transpose of the structure of Figure 4.4 is shown in Figure 4.5.

4.6.2   Infinite Impulse Response (IIR) Filters

Let us consider Equation (4.29) for the case when $N = M = 2$. it is straight forward to extend this analysis to higher orders. The transfer function will reduce to

$$H(z) = \frac{Y(z)}{X(z)} = \frac{b_0 + b_1 z^{-1} + b_2 z^{-2}}{1 + a_1 z^{-1} + a_2 z^{-2}}, \qquad (4.35)$$

which we can write in the following format

$$H(z) = \left(\frac{Y(z)}{W(z)}\right)\left(\frac{W(z)}{X(z)}\right)$$

$$= \left(\frac{b_0 + b_1 z^{-1} + b_2 z^{-2}}{1}\right)\left(\frac{1}{1 + a_1 z^{-1} + a_2 z^{-2}}\right).$$

Let

$$H_1(z) = \frac{W(z)}{X(z)} = b_0 + b_1 z^{-1} + b_2 z^{-2}$$

and

$$H_2(z) = \frac{Y(z)}{W(z)} = \frac{1}{1 + a_1 z^{-1} + a_2 z^{-2}}.$$

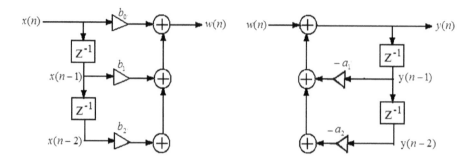

Fig. 4.6 Realization structures for $H_1(z)$ and $H_2(z)$.

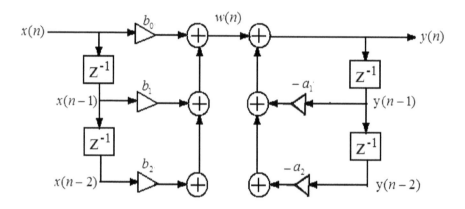

Fig. 4.7 Direct form 1 Realization structure for $H(z)$.

In the time domain we can write two equations

$$w(n) = b_0 x(n) + b_1 x(n-1) + b_2 x(n-2), \quad \text{and}$$

$$y(n) = w(n) - a_1 y(n-1) - a_2 y(n-2).$$

The realization structures formed by the two equations to realize $H_1(z)$ and $H_2(z)$ are shown in Figure 4.6.

Since the first structure generates $w(n)$ which is the input to the second structure the two structures can be joined as in Figure 4.7 to form the realization structure for $H(z)$. Comparing the structure to Equation (4.35) we notice that coefficient of the transfer function $H(z)$ are the multiplier coefficients in the realization structure. Hence

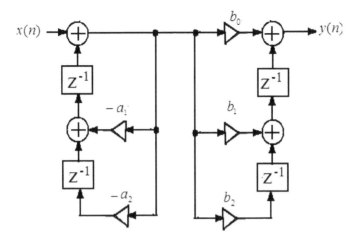

Fig. 4.8 Direct form 1_t.

this structute is a Direct Form 1 structure. We also note there are 4 delays and the order of the transfer function is 2. Hence the structure is noncanonic.

The transpose of the Direct form 1 structure is shown in Figure 4.8.

The Direct form 1_t is also noncanonic. It is possible with further manipulations of these structures to obtain a canonic structure. For instance, in Figure 4.8, moving all delays to be done before the multipli-

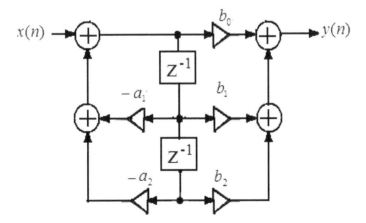

Fig. 4.9 Direct form II structure.

ers it will not change the transfer function. It follows that the delays on the parallel arms have the same inputs and will have the same outputs. These delays can be merged to form the structure which is canonic and is shown in Figure 4.9. This structure is referred to as Direct form II structure. The transpose of Figure 4.9 gives the Direct form II$_t$ structure.

4.6.3   Cascade Realization

Direct form structures of higher order are more sensitive to quantization errors. It is for this reason that a higher order structure is normally implemented as a cascade or parallel connected lower order filters. In cascade realization a high order direct form transfer function like that represented by Equation (4.28) is factored into first- and second-order polynomials.

$$H(z) = \frac{Y(z)}{X(z)} = \frac{\sum_{k=0}^{M} b_k z^{-k}}{\sum_{k=0}^{N} a_k z^{-k}} = C \prod_k \left(\frac{b_{0k} + b_{1k} z^{-1} + b_{2k} z^{-2}}{1 + a_{1k} z^{-1} + a_{2k} z^{-2}} \right).$$
(4.36)

Figure 4.10 shows a cascade of two second-order sections using the direct form II canonic realization. Either of the sections can be reduced to first-order sections using the identity $b_{2k} = a_{2k} = 0$.

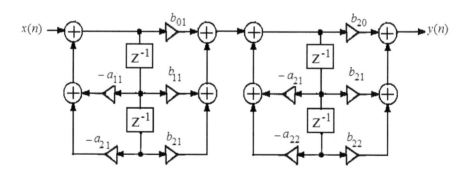

Fig. 4.10 A cascade of two second-order sections.

4.6.4 Parallel Realization

It may also be possible to express Equations (4.29) as a partial fraction expansion. In this case a high order direct form transfer structure is expressed as a parallel realization of first- and second-order sections.

$$H(z) = \frac{Y(z)}{X(z)} = \frac{\sum_{k=0}^{M} b_k z^{-k}}{\sum_{k=0}^{N} a_k z^{-k}}$$

$$= D + \sum_{k} \left(\frac{d_{0k} + d_{1k}z^{-1} + d_{2k}z^{-2}}{1 + c_{1k}z^{-1} + c_{2k}z^{-2}} \right). \tag{4.37}$$

One can obtain first-order sections if $d_{2k} = c_{2k} = 0$ for any section k. In the Figure 4.11, a first-order section is connected in parallel to a second-order section. The transfer function is given as

$$H(z) = \frac{Y(z)}{X(z)}$$

$$= D + \left(\frac{d_{00} + d_{10}z^{-1}}{1 + c_{10}z^{-1}} \right) + \left(\frac{d_{01} + d_{11}z^{-1}}{1 + c_{11}z^{-1} + c_{21}z^{-2}} \right). \tag{4.38}$$

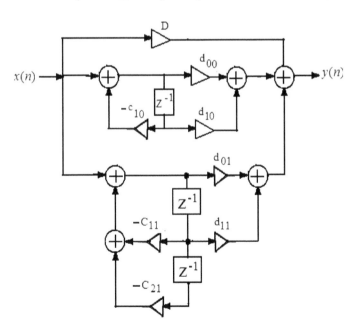

Fig. 4.11 Parallel realization of $H(z)$.

4.7   Problems

4.8.1 Find the z-transforms and the regions of convergence of the following sequences

  (i) $x(n) = (0.5)^n \mu(n - 2)$,

  (ii) $x(n) = -(0.4)^n \mu(-n - 3)$,

  (iii) $x(n) = (-0.4)^n \mu(-n - 3)$.

4.8.2 Derive the z-transform and the region of convergence of the sequence $y(n) = (r^n \cos(\omega_0 n))\mu(n)$. Hence obtain the z-transform of the sequence $w(n) = 2^n \cos\left(\frac{\pi}{2} n\right) \mu(n)$.

4.8.3 Does the following sequence have a z-transform? $w(n) = 2^n \mu(-n - 1) + 3^n \mu(+n + 1)$. Explain your answer.

4.8.4 Using the method of residues obtain the inverse z-transform of $H(z) = \frac{z(z-1)}{(z-2)(z-3)}$ for the following regions of convergence: (i) $|z| > 3$, (ii) $2 < |z| < 3$, and (iii) $|z| < 2$.

4.8.5 Using the method of partial fraction expansion to find the inverse z-transform of $H(z) = \frac{z(z-1)}{(z-2)(z-3)}$ for the following regions of convergence

  (i) $|z| > 3$, (ii) $2 < |z| < 3$, and (iii) $|z| < 2$.

4.8.6 Using the method of partial fraction expansion obtain the inverse z-transform of $H(z) = \frac{(1 - 1/2 z^{-1})}{(1 - 1/3 z^{-1})(1 - 2z^{-1})^2}$ for $|z| > 1/3$.

4.8.7 The transfer function of a causal discrete-time system is given by $H(z) = \frac{z}{3z^2 - 4z + 1}$. Using long division find the first five samples of the impulse response.

4.8.8 The impulse response of an LTI system is given by $h(n) = \alpha_0 \delta(n) + \alpha_1 \delta(n - 1) + \alpha_2 \delta(n - 2) + \alpha_3 \delta(n - 3)$. Obtain the transfer function $H(z)$ of the system.

4.8.9 A moving average filter is a linear time-invariant system whose input/output relationship is given by $y(n) = \frac{1}{M} \sum_{k=0}^{M-1} x(n - k)$. Obtain its impulse response and hence its transfer function $H(z)$.

4.8.10 A sequence is given by $y(n) = ce^{-\alpha n} \mu(n)$, where $\mu(n)$ is a unit step sequence. Making use of table for z-transforms of simple functions obtain the z-transform and the region of convergence of the following functions

(i) $w(n) = ny(n)$, (ii) $v(n) = \beta^n y(n)$, (iii) $u(n) = y(-n)$,
(iv) $s(n) = y(n - n_0)$.

4.8.11 With the aid of a MATLAB program plot the location of poles and zeros for the transfer functions of systems represented by

(i) $H(z) = \frac{z^{-1} - z^{-2}}{(1 - z^{-1} + z^{-2})(1 + 0.8z^{-1})}$

(ii) $3y(n) = 3.7y(n - 1) - 0.7y(n - 2)x(n - 1)$ for $n \geq 0$

Comment on their stability.

4.8.12 A transfer function for an FIR filter is given by $H(z) = (1 - az^{-1} - \beta z^{-2})^3$. Realize the transfer function using

(i) direct form realization, and

(ii) as a cascade of second-order sections.

4.8.13 A transfer function for an IIR filter is given by

$$H(z) = \frac{(1 + 0.5z^{-1})(1 + 0.3z^{-1} + 0.02z^{-2})}{(1 + 0.3z^{-1})(1 + 0.04z^{-1} + 0.2z^{-2})}.$$

Give canonic realization structures using

(a) direct form II realization,

(b) a cascade of a first-order and second-order structures, and

(c) a cascade of two second-order structures.

4.8.14 A Transfer function can be expressed as a partial fraction expansion as follows:

$$H(z) = \gamma_0 + \frac{0.3}{1 + 0.5z^{-1}} + \frac{1 + 0.4z^{-1} + 0.33z^{-2}}{1 - 0.62z^{-1} + 0.43z^{-2}}.$$

Implement a parallel realization structure using not more than second-order structures.

5

Review of Analog Filter Design

5.1 Introduction

The reason for the review of analog filter design before pursuing with digital filter design is twofold. Analog filters are used as anti-aliasing filters and reconstruction filters in digital signal processing. Secondly, a popular design method of IIR filters is through special transformation of analog prototype filters to digital filters. Therefore, it is necessary to be able to specify, design, and implement analog filters.

5.2 Specification of Analog Filters

The essential parameters used to specify an analog lowpass filter are shown in Figure 5.1.

Figure 5.1 represents normalized magnitude response specifications for a lowpass filter. The maximum values of the magnitude in the passband is given by $|H(j\Omega)_{\max}| = 1.0$ and the minimum value of the magnitude in the passband (also referred to the as the passband ripple) is equal to $\frac{1}{\sqrt{1+\varepsilon^2}}$ and is the value at the edge of the passband. The minimum stopband attenuation (also referred to as the maximum stopband ripple) is given as $1/A$. The frequency Ω_p is defined to be the passband edge frequency and the frequency Ω_s is defined to be the stopband edge frequency. The passband is the region where $0 \leq \Omega \leq \Omega_p$, the transition band is the region where $\Omega_p \leq \Omega \leq \Omega_s$ and the stopband is the region where $\Omega_s \leq \Omega \leq \infty$. We also define two more parameters that will help us in the intermediate stages of the design as follows:

(i) The transition (or selectivity) parameter which is defined as the ratio of the passband edge frequency Ω_p and the stopband

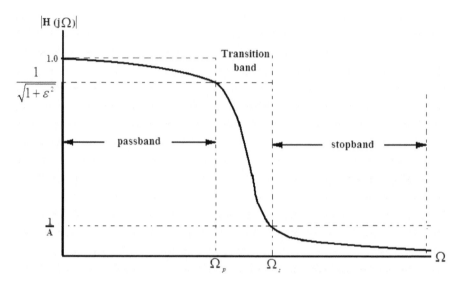

Fig. 5.1 Normalized magnitude response filter specifications.

edge frequency Ω_s, i.e.,

$$k = \frac{\Omega_p}{\Omega_s}. \tag{5.1}$$

(ii) The discrimination parameter which is given by

$$k_1 = \frac{\varepsilon}{\sqrt{A^2 - 1}}. \tag{5.2}$$

5.3   The Analog Lowpass Filters

The main characteristics that define a filter include the passband ripple, the transition band (the cut-off frequency, and the roll-off), the stopband attenuation and the phase response. The different types of filters are designed to optimize one or more of these characteristics. In this review, we will focus only on the four most common types of filters.

5.3.1   Butterworth Filters

The Butterworth filters have a flat magnitude response in the passband characterized by a slow roll-off and a nonlinear phase response. The magnitude square response is given by

$$|H(j\Omega)|^2 = \frac{1}{1 + \left(\frac{\Omega}{\Omega_C}\right)^{2N}}, \tag{5.3}$$

where Ω_C is the cut-off frequency.

The magnitude response is referred to as maximally flat since the first $N - 1$ derivatives of $|H(j\Omega)|^2$ are equal to zero at $\Omega = 0$. The magnitude response in dBs is given by

$$H_{\text{dB}} = 10 \log_{10} |H(j\Omega)|^2 = 10 \log \frac{1}{1 + \left(\frac{\Omega}{\Omega_C}\right)^{2N}} \quad \text{dB}$$

At $\Omega = 0, H_{\text{dB}} = 10 \log 1 = 0$ and at $\Omega = \Omega_C$, $H_{\text{dB}} = 10 \log \frac{1}{1+1^{2N}} = -3\,\text{dB}$.

The frequency Ω_C is referred to as the $3\,\text{dB}$ cut-off frequency. The impact of the order of the filter is more visible in the transition band and in the stopband. To see this we will consider a frequency range for which $\Omega \gg \Omega_C$, where we can approximate the magnitude response as

$$H_{\text{dB}} = 10 \log \frac{1}{\left(\frac{\Omega}{\Omega_C}\right)}^{2N} = -20N \log \left(\frac{\Omega}{\Omega_C}\right).$$

If we consider two frequency points one decade apart $\Omega_2 = 10\Omega_1$, where $\Omega_1 \gg \Omega_C$ then we can write

$$H_{\text{dB2}} = -20N \log \left(\frac{\Omega_2}{\Omega_C}\right) = -20N \log \left(\frac{10\Omega_1}{\Omega_C}\right)$$

$$= -20N \log \left(\frac{\Omega_1}{\Omega_C}\right) - 20N$$

$$= H_{\text{dB1}} - 20N \,\text{dB}. \tag{5.4}$$

From this result we observe that the filter roll-off decreases by $20\,\text{dB}$ whenever the filter order N increases by 1. This is verified in the plots of Figure 5.2, where the magnitude response is plotted for $N = 2, 3$, and 8.

It is clear that the two parameters that are adequate in specifying a Butterworth filter are the cut-off frequency Ω_C and the order of the

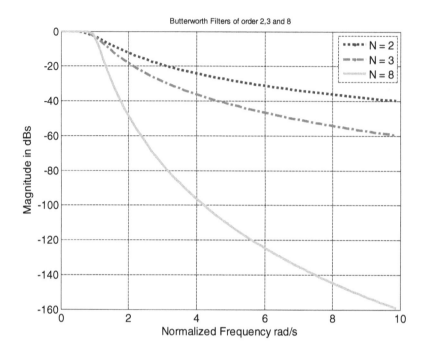

Fig. 5.2 Butterworth filters with different orders.

filter N. In order to compute these two parameters we can use the known values of the magnitude squared response at the passband-edge frequency and the stopband-edge frequency as follows:

At the passband edge frequency

$$|H_C(j\Omega_P)|^2 = \frac{1}{1 + \left(\frac{\Omega_p}{\Omega_C}\right)^{2N}} = \frac{1}{1 + \varepsilon^2}, \qquad (5.5)$$

and at the stopband edge frequency

$$|H_C(j\Omega_P)|^2 = \frac{1}{1 + \left(\frac{\Omega_s}{\Omega_C}\right)^{2N}} = \frac{1}{A^2}. \qquad (5.6)$$

The two equations can be used to solve for the order of the filter giving

$$N = \frac{1}{2}\frac{\log_{10}\frac{(A^2-1)}{\varepsilon^2}}{\log_{10}\frac{\Omega_s}{\Omega_P}} = \frac{\log_{10}\left(\frac{1}{k_1}\right)}{\log_{10}\left(\frac{1}{k}\right)}. \qquad (5.7)$$

Despite the fact that the order of a filter must be an integer the value of N computed from Equation (5.7) is rarely an integer. For this reason the value of N computed from Equation (5.7) must be rounded up to the next higher integer. The integer value of N can then be used in either Equation (5.5) or (5.6) to solve for the cut-off frequency Ω_C. If Equation (5.5) is used then the passband specifications will be met exactly and stopband specifications will be exceeded. On the other hand, if Equation (5.6) is used the stopband specifications will be met exactly and the passband specifications will be exceeded. The decision on which equation to use is left to the designer who may base the decision on the sensitivity of each band to the specific application.

The MATLAB function that uses the above equations to compute the order and the cut-off frequency is buttord and the syntax is given as

$$[N, \Omega_C] = \text{buttord}(\Omega_P, \Omega_s, R_P, R_s, \text{'s'}), \qquad (5.8)$$

where Ω_P is the passband-edge angular frequency in rad/sec,
$\quad\quad\;\; \Omega_s$ is the stopband-edge angular frequency in rad/sec,
$\quad\quad\;\; R_P$ is the maximum passband attenuation in dB,
$\quad\quad\;\; R_s$ is the minimum stopband attenuation in dB, and
$\quad\quad\;\;$'s' must be used to indicate analog filter.

In the implementation of a filter we require the transfer function and its coefficients. The expression of the transfer function of a lowpass Butterworth filter in the s-plane is given by

$$H_a(s) = \frac{1}{D_N(s)} = \frac{\Omega_C^N}{\prod_{i=1}^{N}(s - p_i)}, \qquad (5.9)$$

where the poles are given by $p_i = \Omega_C e^{j\pi(N+2i-1)/2N}$, $i = 1, 2, \ldots, N$. This shows that poles of a lowpass filter are on a circle of radius Ω_C at different phase angles on the left-hand side in the s-plane. Poles have to be on the left-hand side of the imaginary axis in the s-plane for the filter to be stable. The expanded form of the polynomial $D_N(s)$ is tabulated in many text books on circuit theory to simplify the design [6]. The MATLAB function used to compute the transfer function giving the numerator and denominator coefficients is "butter" and the syntax is given as

$$[B, A] = \text{butter}(N, \Omega_C, \text{'type'}, \text{'s'}), \tag{5.10}$$

where B represents a row matrix of numerator coefficients,
$\quad A$ represents a row matrix of denominator coefficients,
$\quad N$ order of the filter, Ω_C is the cut-off frequency in rad/sec,
$\quad$ 'type' specifies the filter type (low, high, bandpss or band-stop), and
$\quad$'s' must be used to indicate analog filter.

Example 5.1. Design and plot the magnitude response of an analog Butterworth lowpass filter to have a maximum passband attenuation of 0.5 dB at 500 Hz and 50 dB minimum stopband attenuation at 1 KHz (see Figure 5.3).

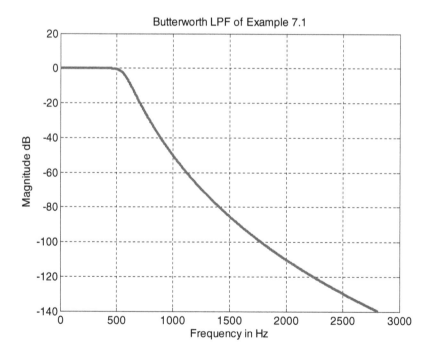

Fig. 5.3 Magnitude response of the Butterworth filter of Example 5.1.

Solution

MATLAB Program

```
% Design Example 5.1
Wp = 500*2*pi;
Ws = 1000*2*pi;
Rp = 0.5;
Rs = 50;
[N, Wc] = buttord(Wp, Ws, Rp, Rs, 's');
[B, A] = butter(N, Wc, 's');
omega= [0:1:5*Wc];
h = freqs(B,A,omega);
plot(omega./(2*pi),20*log10(abs(h)),'m');
title('Butterworth LPF Example 5.1');
xlabel('Frequency in Hz'); ylabel('Magnitude');
```

5.3.2 Chebyshev Filters

The Chebyshev filter magnitude response has a much more steeper roll-off at a penalty of an amount of ripples in the passband or stopband, depending on the type of Chebyshev transfer function. There are two type of Chebyshev transfer functions; type 1 Chebyshev transfer functions are equiripple in the passband and monotonic in the stopband while type II are monotonic in the passband and equiripple in the stopband.

5.3.2.1 Chebyshev Type I Transfer Function

The squared magnitude response of the analog type I Chebyshev filter is given by

$$|H_a(j\Omega)|^2 = \frac{1}{1 + \varepsilon^2 T_N^2\left(\frac{\Omega}{\Omega_p}\right)}, \tag{5.11}$$

where $T_N(x)$ is the Chebyshev polynomial of order N which is given by

$$T_N(x) = \begin{cases} \cos(N\cos^{-1}x) & |x| \leq 1 \\ \cosh(N\cosh^{-1}x) & |x| > 1 \end{cases}. \tag{5.12}$$

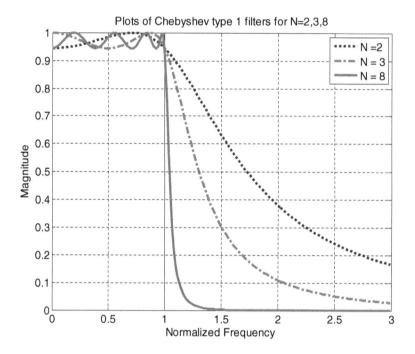

Fig. 5.4 Chebyshev type I filters with different orders.

The magnitude responses for different type I Chebyshev filters are given in Figure 5.4 for $N = 2$, 3 and 8.

Figure 5.4 shows that the filter is equiripple in the passband, i.e., $0 \leq \Omega \leq 1$, and monotonic for $\Omega > 1$. The roll-off is also much more than for the Butterworth filter.

The order N of the filter can be determined from Equation (5.11) using the attenuation specifications as follows:

$$|H_a(j\Omega)|^2 = \frac{1}{1 + \varepsilon^2 T_N^2\left(\frac{\Omega_s}{\Omega_p}\right)} = \frac{1}{A^2},$$

which after substitution using (5.12) and solving for N we get

$$N = \frac{\cosh^{-1}\frac{\left(\sqrt{A^2-1}\right)}{\varepsilon}}{\cosh^{-1}\left(\frac{\Omega_s}{\Omega_P}\right)} = \frac{\cosh^{-1}\left(\frac{1}{k_1}\right)}{\cosh^{-1}\left(\frac{1}{k}\right)}. \tag{5.13}$$

The MATLAB function that uses Equation (5.13) to compute the order and the cut-off frequency is cheb1ord and the syntax is given as

$$[N, \Omega_C] = \text{cheb1ord}(\Omega_P, \Omega_s, R_P, R_s, \text{'}s\text{'}), \qquad (5.14)$$

where Ω_P is the passband edge angular frequency in rad/sec,
$\quad \Omega_s$ is the stopband edge angular frequency in rad/sec,
$\quad R_P$ is the maximum passband attenuation in dB,
$\quad R_s$ is the minimum stopband attenuation in dB, and
$\quad$'s' must be used to indicate analog filter.

The transfer function of the Chebyshev type I filter is of the same form as that of the Butterworth filter given by Equation (5.9) and can be expressed as a ratio of two polynomials. The coefficients of the rational transfer function can be obtained by the MATLAB function cheby1 and the syntax is given by

$$[B, A] = \text{cheby1}(N, R_P, \Omega_C, \text{'}s\text{'}), \qquad (5.15)$$

where B represents a row matrix of numerator coefficients,
$\quad A$ represents a row matrix of denominator coefficients,
$\quad N$ order of the filter, Ω_C is the cut-off frequency in rad/sec, and
$\quad$'s' must be used to indicate analog filter.

Example 5.2. Design and plot the magnitude response of an analog Chebyshev type I lowpass filter to have a maximum passband attenuation of 0.5 dB at 500 Hz and 50 dB minimum stopband attenuation at 1 KHz (see Figure 5.5).

Solution

MATLAB Program

```
% Design Example 5.2
Wp = 500*2*pi;
Ws = 1000*2*pi;
Rp = 0.5;
```

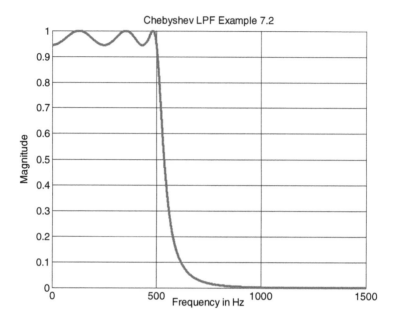

Fig. 5.5 Magnitude response of the Chebyshev type I filter of Example 5.2.

```
Rs = 50;
[N, Wc] = cheb1ord(Wp, Ws, Rp, Rs, 's');
[B, A] = cheby1(N, Rp, Wc, 's');
omega = [0:1:3*Wc];
h = freqs(B,A,omega);
plot(omega./(2*pi),(abs(h)),'m');
title('Chebyshev LPF Example 5.2');
xlabel('Frequency in Hz'); ylabel('Magnitude');
```

5.3.2.2   Chebyshev Type II Transfer Function

The magnitude-squared response for the type II Chebyshev transfer function is given by

$$|H_a(j\Omega)|^2 = \frac{1}{1 + \varepsilon^2 \left[\frac{T_N(\gamma)}{T_N(\lambda)}\right]^2}, \qquad (5.16)$$

where $\gamma = \frac{\Omega_s}{\Omega_P}$ and $\lambda = \frac{\Omega_s}{\Omega}$. The magnitude-squared response is flat in the passband and equiripple in the stopband. Equation (5.13) can be used to compute the value of N. The same equation is used in the m file cheb2ord of MATLAB to compute the order N and the cut-off frequency W_C as follows:

$$[N, W_C] = \text{cheb2ord}(W_p, W_s, R_P R_s, 's'). \tag{5.17}$$

The transfer function $H_a(s)$ can be written as a ratio of two polynomials. In order to compute the coefficients the MATLAB function Cheb2 can be used as follows:

$$[B, A] = \text{cheb2}(N, R_s, W_C, 'type', 's'). \tag{5.18}$$

Example 5.3. Design and plot the magnitude response of an analog Chebyshev type II lowpass filter to have a maximum passband attenuation of 0.5 dB at $500\,\text{Hz}$ and 50 dB minimum stopband attenuation at $1\,\text{KHz}$ (see Figure 5.6).

Solution

MATLAB Program

```
% Design Example 5.3
Wp = 500*2*pi;
Ws = 1000*2*pi;
Rp = 0.5;
Rs = 50;
[N, Wc] = cheb2ord(Wp, Ws, Rp, Rs, 's');
[B, A] = cheby2(N, Rs, Wc, 's');
omega = [0:1:3*Wc];
h = freqs(B,A,omega);
plot(omega./(2*pi),20*log10(abs(h)),'m');
title('Chebyshev Type II LPF Example 5.3');
xlabel('Frequency in Hz'); ylabel('Magnitude');
```

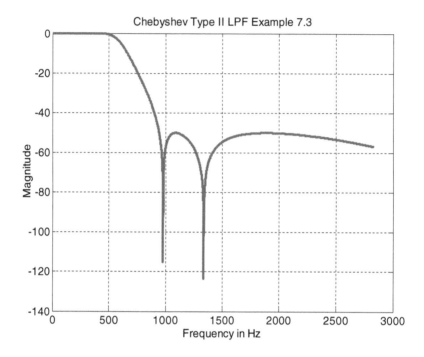

Fig. 5.6 Magnitude response of the Chebyshev type II filter of Example 5.3.

5.3.3   The Elliptic Filters

The elliptic filter transfer function is equiripple both in the passband and in the stopband. The amount of ripple in each band can be independently adjusted. The elliptic filter magnitude response has the steepest roll-off compared to any other type of filter of the same order. The analysis is complicated [7] and here we only make reference to some interesting properties.

The magnitude-squared response is given by

$$H_a(j\Omega) = \frac{1}{1 + \varepsilon^2 R_N^2\left(\xi, \frac{\Omega}{\Omega_0}\right)}, \qquad (5.19)$$

where R_N is the nth order elliptic rational function, Ω_0 is the cut-off frequency, ε is the ripple factor and ξ is the selectivity factor. Whereas ε defines the ripple in the passband, the combination of ε and ξ defines

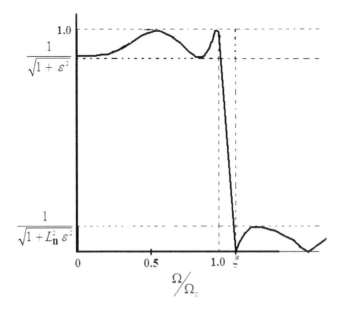

Fig. 5.7 Magnitude response of an elliptic filter.

the ripple in the stopband. In Figure 5.7

$$L_N = R_N \left(\xi, \frac{\Omega}{\Omega_0} \right).$$

It is interesting to note that as $\xi \to \infty$, the rational function R_N becomes a chebyshev polynomial and the magnitude response becomes a Chebyshev type I response. Also as $\xi \to \infty$, $\varepsilon \to 0$, and $\Omega_0 \to 0$ such that $\xi \Omega_0 = 1$ and $\varepsilon L_n = \alpha$ (constant), the magnitude response becomes that of a Chebyshev type II.

In order to find the order we use the approximation from [9] which requires following specifications: the passband edge frequency Ω_p, the stopband edge frequency Ω_s, and the passband ripple ε. The approximation is given by

$$N = \frac{2 \log_{10} \left(\frac{4}{k_1} \right)}{\log \left(\frac{1}{\rho} \right)}, \tag{5.20}$$

where $k_1 = \frac{\varepsilon}{\sqrt{A^2-1}}$ the selectivity parameter,

$$\rho = \rho_0 + 2(\rho_0)^5 + 15(\rho_0)^9 + 150(\rho_0)^{13},$$

$$k = \frac{\Omega_p}{\Omega_s}, \quad k' = \sqrt{1-k^2}, \quad \text{and} \quad \rho_0 = \frac{1-\sqrt{k'}}{2(1+\sqrt{k'})}.$$

The MATLAB m-file function used to compute the order of the lowpass filter is

$$[N, W_C] = \text{ellipord}(\Omega_P, \Omega_s, R_P, R_s, 's'). \tag{5.21}$$

In order to obtain the coefficients of the transfer function for the lowpass filter we use

$$[B, A] = \text{ellip}(N, R_P, R_s, W_C). \tag{5.22}$$

Example 5.4. Design and plot the magnitude response of an analog elliptic lowpass filter to have a maximum passband attenuation of 0.5 dB at 500 Hz and 50 dB minimum stopband attenuation at 550 Hz (see Figure 5.8).

MATLAB Program

```
% Design Example 5.4 Elliptic Lowpass Filter
Wp = 500*2*pi;
Ws = 550*2*pi;
Rp = 0.5;
Rs = 50;
[N, Wc] = ellipord(Wp, Ws, Rp, Rs, 's');
[B, A] = ellip(N, Rp, Rs, Wc, 's');
omega = [0:1:3*Wc];
h = freqs(B,A,omega);
plot(omega./(2*pi),20*log10(abs(h)),'m');
title('Elliptic LPF Example 5.4');
xlabel('Frequency in Hz'); ylabel('Magnitude');
```

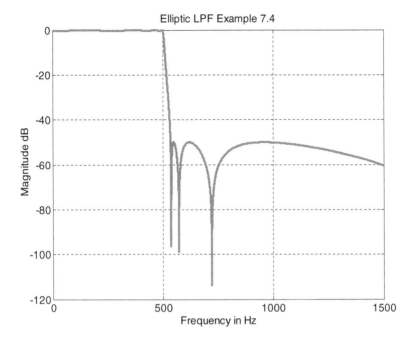

Fig. 5.8 Magnitude response for the Elliptic filter of Example 5.4.

5.3.4 The Bessel Filters

The four types of filter we have discussed so far are designed to meet certain magnitude against frequency specifications. The phase has been totally ignored. We have mentioned previously that the phase-frequency characteristic in the passband is nonlinear for these filters and the extent of nonlinearity is different with each one of them. Figure 5.9 shows a comparison of the phase nonlinearity for the Butterworth, Chebyshev 1, Chebyshev II, and the elliptic filters that are of the same order. The black line at the moment is the reference line for a perfectly linear filter. The phase discontinuity due to the plotting range being limited to $-\pi \leq \theta \leq \pi$ has been removed by the function unwrap. It can be observed that all the four filters have a significant degree of nonlinearity in the phase frequency characteristics. The Butterworth filter has the least nonlinearity followed by Chebyshev I, Chebyshev II, and the elliptic filter has the most nonlinearity.

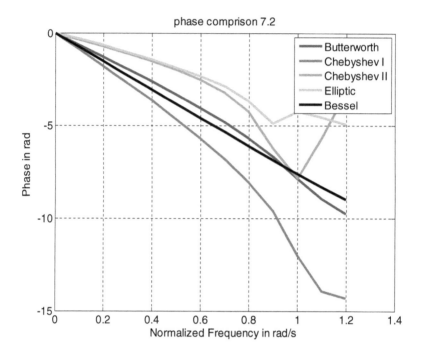

Fig. 5.9 Comparison of Phase-frequency characteristics for different filter types.

The Bessel filters have a linear phase-frequency characteristic. The transfer function is given by

$$H(s) = \frac{\theta_N(0)}{\theta_N(s/\Omega_0)} \quad \text{where } \theta_N(x) = \sum_{k=0}^{N} \frac{(N+k)!}{(N-k)!k!} \left(\frac{x}{2}\right)^k. \quad (5.23)$$

For $N = 3$ $H(s) = \frac{1}{1+6x+15x^2+15x^3}$ where $x = s/\Omega_0$.

Notice that $H(s)$ has only poles and no zeros. In order to determine the numerator and denominator of the transfer function we use MATLAB function

$$[B, A] = \text{besself}(N, W_c).$$

The black line in Figure 5.9 is in fact the plot for phase-frequency characteristics of the Bessel filter and is perfectly linear.

The penalty for perfect linearity in the phase-frequency characteristic is in the magnitude response. This can be seen in the plots of

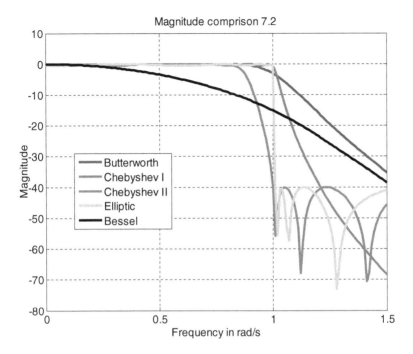

Fig. 5.10 Magnitude response for various types of filters.

Figure 5.10. The Bessel filter has the slowest transition from the pass-band to the stopband, followed by the Butterworth filter. The elliptic filter has the sharpest roll-off in the transition region compared to the rest of the filter.

5.4 The Analog Highpass, Bandpass, and Bandstop Filters

Other types of analog filters like the highpass, bandpass, and bandstop filters can be designed from the lowpass filter by means of spectral transformation. The spectral transformation will maintain the basic characteristics of the lowpass filter in the transformed filter. If the original lowpass filter is a Butterworth filter the transformed filter will also be a Butterworth filter and so on. The derivations of these spectral transformations have been done in [10]. In this section, we will only revise the design process in a flow chart.

5.4.1 Design Procedure for a Highpass Filter

The following variables are used in the design flow Chart 5.1:

Ω_P = passband-edge frequency in the prototype LPF

Ω_s = stopband-edge frequency in the prototype LPF

$\hat{\Omega}_p$ = passband-edge frequency in the desired HPF

$\hat{\Omega}_s$ = stopband-edge frequency in the desired HPF

R_P = Maximum passband attenuation in dB

R_s = Minimum stopband attenuation in dB

B = Row Matrix representing Numerator Coefficients of prototype LPF

A = Row Matrix representing Denominator Coefficients of prototype LPF

Num = Row Matrix representing Numerator Coefficients of desired HPF

Den = Row Matrix representing Denominator Coefficients of desired HPF

5.4.2 Design Procedure for a Bandpass Filter (BPF)

The following variables are used in the flow Chart 5.2:

Ω_P = passband-edge frequency in the prototype LPF

Ω_s = stopband-edge frequency in the prototype LPF

$\hat{\Omega}_{p1}$ = lower passband-edge frequency in the desired BPF

$\hat{\Omega}_{p2}$ = higher passband-edge frequency in the desired BPF

$\hat{\Omega}_{s1}$ = lower stopband-edge frequency in the desired BPF

$\hat{\Omega}_{s1}$ = higher stopband-edge frequency in the desired BPF

R_P = Maximum passband attenuation in dB

R_S = Minimum stopband attenuation in dB

B = Row Matrix representing Numerator Coefficients of prototype LPF

A = Row Matrix representing Denominator Coefficients of prototype LPF

Num = Row Matrix representing Numerator Coefficients of desired BPF

Den = Row Matrix representing Denominator Coefficients of desired BPF

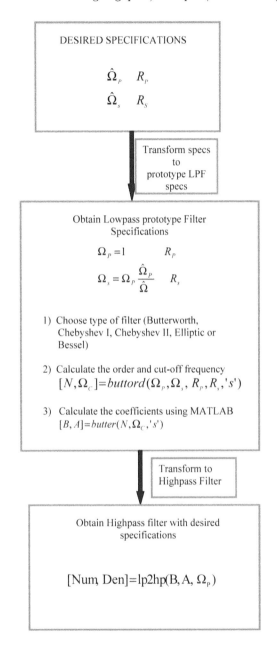

Chart 5.1 Flow Chart for the design procedure for analog highpass filters.

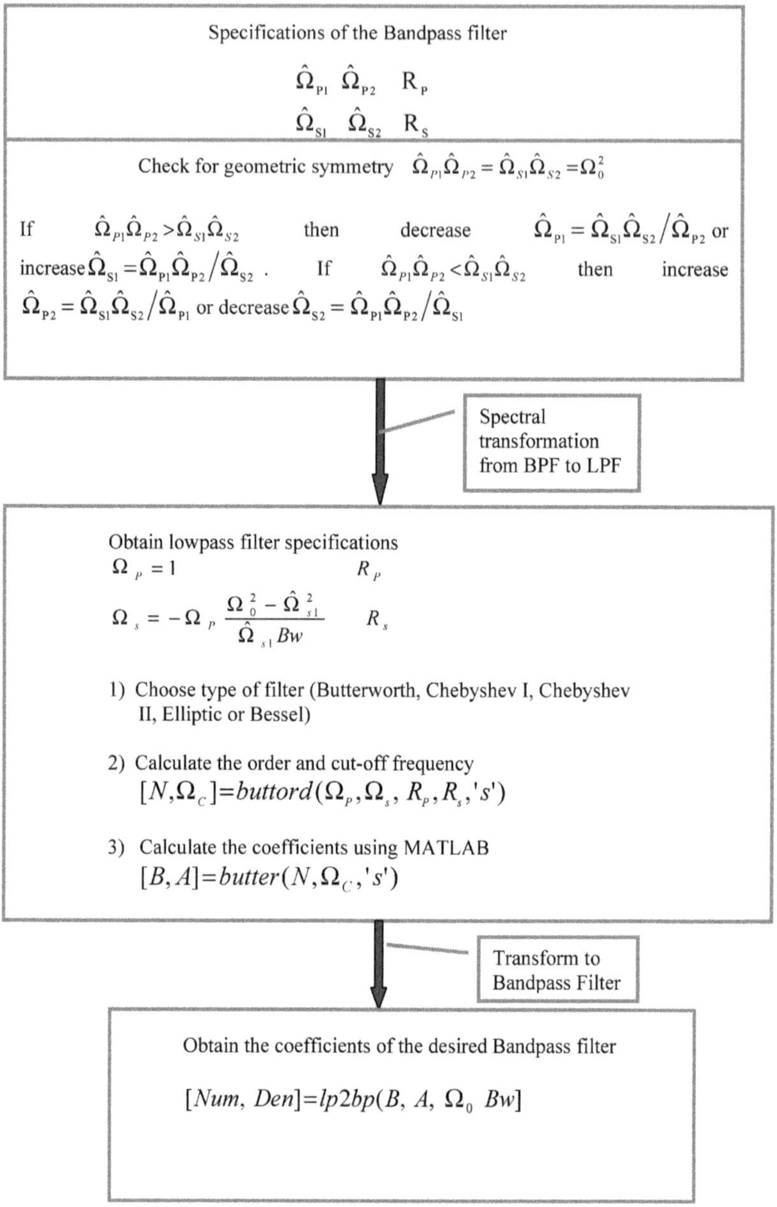

Specifications of the Bandpass filter

$$\hat{\Omega}_{P1} \quad \hat{\Omega}_{P2} \quad R_P$$
$$\hat{\Omega}_{S1} \quad \hat{\Omega}_{S2} \quad R_S$$

Check for geometric symmetry $\hat{\Omega}_{P1}\hat{\Omega}_{P2} = \hat{\Omega}_{S1}\hat{\Omega}_{S2} = \Omega_0^2$

If $\quad \hat{\Omega}_{P1}\hat{\Omega}_{P2} > \hat{\Omega}_{S1}\hat{\Omega}_{S2} \quad$ then $\quad$ decrease $\quad \hat{\Omega}_{P1} = \hat{\Omega}_{S1}\hat{\Omega}_{S2}/\hat{\Omega}_{P2}$ or increase $\hat{\Omega}_{S1} = \hat{\Omega}_{P1}\hat{\Omega}_{P2}/\hat{\Omega}_{S2}$. If $\quad \hat{\Omega}_{P1}\hat{\Omega}_{P2} < \hat{\Omega}_{S1}\hat{\Omega}_{S2} \quad$ then $\quad$ increase $\hat{\Omega}_{P2} = \hat{\Omega}_{S1}\hat{\Omega}_{S2}/\hat{\Omega}_{P1}$ or decrease $\hat{\Omega}_{S2} = \hat{\Omega}_{P1}\hat{\Omega}_{P2}/\hat{\Omega}_{S1}$

Spectral transformation from BPF to LPF

Obtain lowpass filter specifications

$$\Omega_p = 1 \qquad\qquad R_P$$

$$\Omega_s = -\Omega_p \frac{\Omega_0^2 - \hat{\Omega}_{s1}^2}{\hat{\Omega}_{s1} Bw} \qquad R_s$$

1) Choose type of filter (Butterworth, Chebyshev I, Chebyshev II, Elliptic or Bessel)

2) Calculate the order and cut-off frequency
$$[N,\Omega_C] = buttord(\Omega_p, \Omega_s, R_p, R_s, 's')$$

3) Calculate the coefficients using MATLAB
$$[B, A] = butter(N, \Omega_C, 's')$$

Transform to Bandpass Filter

Obtain the coefficients of the desired Bandpass filter

$$[Num, Den] = lp2bp(B, A, \Omega_0 \; Bw]$$

Chart 5.2 Flow chart for the design procedure for analog bandpass filters.

5.4.3 Design Procedure for a Bandstop Filter (BSF)

The Design procedure for the Bandstop filter is similar to the design procedure for the bandstop filter with the following differences

(1) For the lowpass prototype filter $\Omega_S = 1$ and $\Omega_P = \Omega_S \frac{\hat{\Omega}_P Bw}{\hat{\Omega}_0^2 - \hat{\Omega}_P^2}$.
(2) $Bw = \hat{\Omega}_{s2} - \hat{\Omega}_{s1}$.
(3) The coefficients of the transfer function are obtained from [Num, Den] $= lp2bs(B, A, \Omega_0 Bw)$.

At the end of this section a program to design a bandstop filter is given.

Example 5.5. An elliptic analog bandstop filter is to be designed to eliminate an interfering radio armature transmission at 144 MHz. The filter has the following specifications:

> Passband edges: 115 MHz and 175 MHz
> Stopband edges: 125 MHz and 165 MHz
> Peak passband ripple is to be 0.5 dB
> Minimum stopband attenuation is to be 30 dB

Determine the order of the prototype lowpass filter to be used in the design. Determine the order of the bandstop filter and its transfer function. Plot the Magnitude response of the lowpass filter and the bandstop filter.

Solution

MATLAB Program

```
% This programme will design an Analog elliptic
  bandstop filter
Wp1 = input('the lower passband-edge frequency in Hz,
       Wp1 = ');
Wp2 = input('the higher passband-edge frequency in Hz,
       Wp2 = ');
Ws1 = input('the lower stopband-edge frequency in Hz,
       Ws1 = ');
```

```
Ws2 = input('the higher stopband-edge frequency in Hz,
        Ws2 = ');
Rp = input('the maximum passband attenuation in dBs,
        Rp = ');
Rs = input('the minimum stopband attenuation in dBs,
        Rs = ');
% Check for geometric symmetry
Wp1=Wp1*2*pi;
Wp2=Wp2*2*pi;
Ws1=Ws1*2*pi;
Ws2=Ws2*2*pi;
if Wp1*Wp2$>$Ws1*Ws2
    Wp1=Ws1*Ws2/Wp2;
    WoSquared =Ws1*Ws2;
else Ws2 = Wp1*Wp2/Ws1;
    WoSquared = Wp1*Wp2;
end
% Specify the prototype LPF
Bw = Ws2 - Ws1;
Omegas =1;
Wo = sqrt(WoSquared);
Omegap = Omegas*(Wp1*Bw)/(WoSquared - Wp1^2);
disp(Omegap);
% Obtain the order of the prototype LPF
[N,Wc]= ellipord(Omegap, Omegas, Rp, Rs, 's');
disp('Order of the lowpass filter'); disp(N);
% Obtain the coefficients of the prototype LPF
[B,A] = ellip(N, Rp, Rs, Wc, 's');
% Write the transfer function of the lowpass filter
disp('The transfer function of the LPF');
hlpf = tf(B,A)
% Make a spectral transformation to the bandstop filter
[Num, Den] = lp2bs(B, A, Wo, Bw);
disp('The transfer function of the LPF');
hbsf = tf(Num,Den)
% Plot the Magnitude response of the LPF
```

```
Omega1 = [0: 0.01: 3*Wc];
HLP = freqs(B,A,Omega1);
subplot(2,1,1)
plot(Omega1,abs(HLP),'m');
title('Magnitude Response of the LP prototype Filter');
xlabel('Frequency, Hz');ylabel('Magnitude');
% Plot the Magnitude response of the Bandpass filter
Omega2 = 2*pi*[0: Wo/100: Wo];
HBP = freqs(Num,Den,Omega2);
subplot(2,1,2)
plot(Omega2/(2*pi), abs(HBP));
title('Magnitude Response of the Bandstop Filter');
xlabel('Frequency, Hz');ylabel('Magnitude');
```

Results

    Order of the lowpass filter 4

    Order of the bandstop filter 8

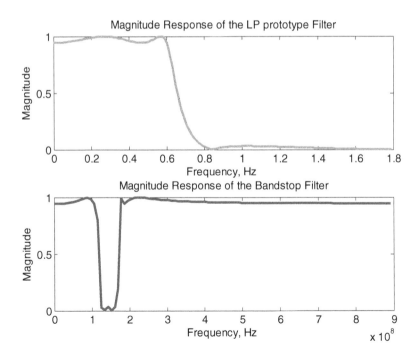

5.5   Problems

5.5.1 An analog Butterworth lowpass filter is to be designed to meet the following specifications:

Passband-edge frequency $\Omega_P = 2\pi\,500\,\text{rad/s}$
Stopband-edge frequency $\Omega_P = 2\pi\,1250\,\text{rad/s}$
Maximum passband attenuation $= 1\,\text{dB}$
Minimum stopband attenuation $= 30\,\text{dB}$
Determine the following

(i) the transition parameter,

(ii) the discrimination parameter, and

(iii) the order of the filter.

5.5.2 A second-order Butterworth lowpass filter with a cut-off frequency of $1\,\text{kHz}$ is to be designed. Obtain the transfer function of the filter and the pole positions on the s-plane.

5.5.3 An analog Chebyshev I lowpass filter is to be designed to meet the following specifications:

Passband-edge frequency $\Omega_P = 2\pi\,500\,\text{rad/s}$
Stopband-edge frequency $\Omega_P = 2\pi\,750\,\text{rad/s}$
Maximum passband attenuation $= 1\,\text{dB}$
Minimum stopband attenuation $= 40\,\text{dB}$
Determine the following

(iv) the transition parameter,

(v) the discrimination parameter, and

(vi) the order of the filter.

Use MATLAB functions to verify the order of the filter. Obtain the transfer function and plot the magnitude and phase response of the filter.

5.5.4 Repeat Problem 5.5.3 if the desired filter is an elliptic filter.

5.5.5 A stage in a student project requires a highpass analog filter that will remove the bass component of music from a CD. Any music component below $500\,\text{kHz}$ is to be attenuated by at least $50\,\text{dBs}$ and any music component above $550\,\text{kHz}$ is not supposed to be attenuated by more than 0.5 dB. The student wanted to use a filter with the lowest order and for that reason he selected an elliptic filter. Determine the following:

(i) The specifications of the lowpass prototype filter used in the intermediate stage of the design.

(ii) The order of the prototype filter and its transfer function.

(iii) The order of the highpass filter and its transfer function.

(iv) Plot the magnitude and phase response of the prototype and the Highpass filters.

(v) Comment on the suitability of the phase response from this filter for the particular application.

5.5.6 A student who is designing an analog AM radio receiver has to down-covert the radio-frequency $(r\text{-}f)$ signal to an intermediate frequency $(i\text{-}f)$ signal. The $i\text{-}f$ stage consists of bandpass filter centered around 455 kHz and has a bandwidth of 10 kHz. Within the passband the $i\text{-}f$ signal should not be attenuated by more than 1 dB. To ensure that adjacent channel interference is reduced to a minimum the carrier at the adjacent channel must be attenuated by at least 40 dB. If the channel spacing is 10 kHz design the filter that will give adequate selectivity. Give the order of the filter, the transfer function and plot the magnitude response if the student used Chebyshev type II filter in his design.

5.5.7 The spectrum of an ECG signals spans the frequency range from DC to 130 Hz. There is an interfering signal from a 50 Hz power-line. A student makes an attempt to suppress this signal using a narrowband bandstop filter centered around 50 Hz with a bandwidth of 5 Hz (i.e., the width between stopband-edge frequencies). Any signal falling within the stopband should be attenuated by at least 50 dB and any signal falling within the passband should not be attenuated by more than 0.5 dB. The width between the passband edge should be about 10 Hz. Design the appropriate filter giving the filter order, the transfer function and plot of the magnitude response of the final filter. Choose a filter-type that will give you the lowest order and steepest roll-off.

6

Digital Filter Design

6.1 Introduction

The objective of filter design is to find a stable function that is realizable using a suitable filter structure to estimate a specified frequency response or impulse response. There are two classes of filters based on the length of their impulse response. The first class includes any filter whose impulse response is of finite length. Such filters are referred to as Finite Impulse Response (FIR) digital filters. These filters are always stable and can be designed to have exactly linear phase. The second class of filters includes any filter whose impulse response is of infinite length. Such filters are referred to as Infinite Impulse Response (IIR) digital filters. These filters are not always stable and cannot be designed to have exactly linear phase. However, IIR filters have one significant advantage over FIR filters. It is possible to approximate a specified magnitude response using an IIR filter of a much lower order than that of an FIR filter. This means that the computational complexity in the implementation of an IIR filter is much less than that of an FIR filter in order to achieve the same objectives.

The design procedure requires that the specifications of the filter are developed from the intended application. The specifications must be given such that the filter will introduce minimum distortion in the desired bands of the signal. The ideal filter will have a gain of unity in the passband and zero in the stopband with the transition bandwidth that is zero. Such a filter is not realizable in practice as it is unstable and noncausal. In order to get a stable and realizable filter the specifications are relaxed to allow some tolerance in the passband and stopband and a gradual transition from the passband to the stopband. The resulting specifications are very similar to those for the analog

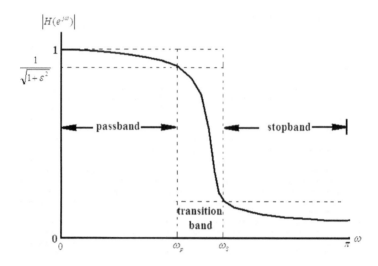

Fig. 6.1 Magnitude specifications for a digital LPF.

filters except it must be taken into account that the magnitude response
of a discrete-time system is periodic. The normalized specifications
are shown in Figure 6.1. The specification parameters are defined as
follows:

The maximum passband deviation $= \frac{1}{\sqrt{1+\varepsilon^2}}$

The maximum passband attenuation $= \alpha_{\max} = 20\log(\sqrt{1+\varepsilon^2})$

The maximum stopband magnitude $= 1/A$

The passband-edge frequency in radians/sample $= \omega_p$

The stopband-edge frequency in radians/sample $= \omega_s$.

It should be noted that frequencies are specified in Hz but digital filter are specified using normalized angular frequencies in units of radians
per sample. The normalization is done by dividing the analog frequency
by the sampling frequency F_S. The analog passband edge frequencies
Ω_P and Ω_S, for instance, are normalized to give $\omega_P = \frac{\Omega_P}{F_S} = 2\pi\left(\frac{f_P}{F_S}\right)$,
and $\omega_S = \frac{\Omega_S}{F_S} = 2\pi\left(\frac{f_S}{F_S}\right)$.

The design methods for the two classes of filters are very different.
In Sections 6.2 and 6.3 we review the most common design methods
for the two classes of filters.

6.2   IIR Filter Design

IIR filters are designed by a mapping process from the s-plane to the z-plane. Initially the specifications of the digital filter are transformed into specifications of an analog filter in the s-plane. An analog filter is then designed before it is transformed back to a digital filter in the z-plane. The advantage of designing an analog filter instead of a digital filter directly is mainly to make use of the mature and advanced knowledge already available in the design of analog filters. The important requirement in the mapping process is to make sure that essential properties of the analog filters are maintained. The imaginary axis in the s-plane must be mapped into the unit circle on the z-plane and a stable analog filter must be mapped to a stable digital filter.

There are two common approaches used in IIR filter design; the impulse invariance method and bilinear z-transformation method. The impulse invariance method preserves the impulse response of the analog filter as the impulse response of the digital filter is formed by sampling the impulse response of the analog filter. However, the magnitude response is not preserved due to aliasing. The impulse invariance method is therefore not a good method for the design of highpass, bandpass, and bandstop digital filters. On the other hand, the bilinear z-transform method is suitable for preserving the magnitude response but not the impulse response. It is therefore suitable for the design of frequency selective filters. In Section 6.2, we will focus on the bilinear z-transform method.

6.2.1   The Bilinear Transformation Method

The bilinear transformation is a mapping of points between the z-plane and the s-plane and is given by the relation [1]

$$s = \frac{2}{T}\left(\frac{1 - z^{-1}}{1 + z^{-1}}\right), \quad \text{and} \tag{6.1a}$$

the reverse transformation is given by

$$z = \frac{1 + \left(\frac{T}{2}\right)s}{1 - \left(\frac{T}{2}\right)s}. \tag{6.1b}$$

This transformation represents a one-to-one mapping where points in the s-plane are uniquely mapped onto points in the z-plane and vice versa. The procedure for designing the IIR digital filter involves a two way transformation. The digital filter specifications are transformed into the analog lowpass filter specifications and this is followed by the design of the analog lowpass filter. The transfer function of the analog lowpass filter is then inverse-transformed into the transfer function of the digital filter. In the process of transformation and inverse transformation the factor $2/T$ is canceled out. Thus it is sufficient to represent the mapping with simply the following transformations

$$s = \left(\frac{1 - z^{-1}}{1 + z^{-1}} \right), \quad \text{and} \tag{6.2a}$$

the reverse transformation is given by

$$z = \frac{1 + s}{1 - s}. \tag{6.2b}$$

A digital filter transfer function $H(z)$ can therefore be obtained from the analog filter with a transfer function $H_a(s)$ as

$$H(z) = H_a(s)|_{\frac{1-z^{-1}}{1+z^{-1}}}. \tag{6.3}$$

In order to determine whether the important properties of the analog filter are retained by the digital filter after the transformation we investigate the mapping of the various regions of the s-plane into the z-plane. If we substitute $s = \sigma + j\Omega$ into the expression for $|z|^2$ obtained from Equation (6.2b) we get

$$|z|^2 = \frac{|1 + \sigma + j\Omega|^2}{|1 - \sigma - j\Omega|^2} = \frac{(1 + \sigma)^2 + \Omega^2}{(1 - \sigma)^2 + \Omega^2}. \tag{6.4}$$

Consider the following cases from Equation (6.4):

(i) When $\sigma < 0$, $|z| < 1$

Points that are to the left of the imaginary axis in the s-plane are mapped interior to the unit circle in the z-plane. A stable analog filter transfer function with poles on the left-hand side of the imaginary axis on s-plane is mapped into a stable digital filter with poles inside the unit circle in the z-plane.

(ii) When $\sigma = 0$, $|z| = 1$

When $\sigma=0$ then $s = j\Omega$. The imaginary axis is mapped onto the unit circle.

(iii) When $\sigma > 0$, $|z| > 1$

An unstable analog filter transfer function with one or more poles on the right-hand side of the imaginary axis in the s-plane is mapped into an unstable digital filter transfer function with poles exterior to the unit circle in the z-plane.

We have seen that the imaginary axis in the s-plane represented by $s = j\Omega$ is mapped into the unit circle represented by $z = e^{j\omega}$ in the z-plane. If we use Equation (6.2a) and substitute $s = j\Omega$ and $z = e^{j\omega}$ we get the following:

$$s = \left(\frac{1 - z^{-1}}{1 + z^{-1}}\right) \quad \text{gives}$$

$$j\Omega = \frac{1 - e^{-j\omega}}{1 + e^{-j\omega}} = \frac{e^{\frac{-j\omega}{2}}}{e^{\frac{-j\omega}{2}}} \left(\frac{e^{\frac{-j\omega}{2}} - e^{\frac{-j\omega}{2}}}{e^{\frac{-j\omega}{2}} + e^{\frac{-j\omega}{2}}}\right)$$

$$\Omega = \tan\left(\frac{\Omega}{2}\right) \quad \text{or}$$

$$\omega = 2\tan^{-1}\Omega. \tag{6.5}$$

From Figure 6.2 it can be observed that the transformation is linear in two locations with very $-0.5 < \Omega < 0.5$; just before and just after the origin and hence the name bilinear transformation. Beyond the linear location the transformation is highly nonlinear. For instance one can observe that the whole of the positive $j\Omega$ axis ($0 < \Omega < \infty$) is mapped into a range of digital frequencies Ω from 0 to π. Through the mapping the analog frequency range is nonlinearly compressed to a much smaller range. The mapping therefore introduces a frequency distortion referred to as frequency warping. The effect of frequency warping on a frequency magnitude response is shown in Figure 6.3.

In order to cancel the distortion due to frequency warping the band-edge frequencies must be pre-warped using the relation $\Omega = \tan(\omega/2)$. The analog filter design is then made with the pre-warped analog angular frequencies. The distortion will be canceled when the analog

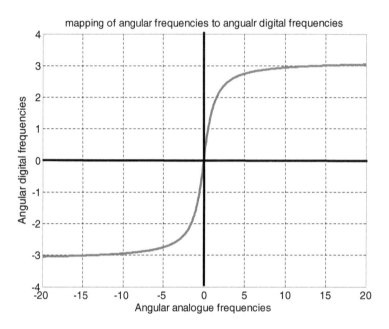

Fig. 6.2 Mapping of angular analog frequencies to analog digital frequencies $\omega = 2\tan^{-1}\Omega$.

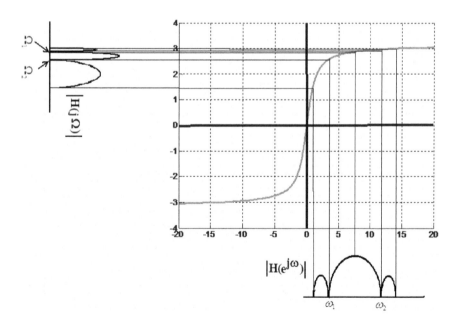

Fig. 6.3 Frequency warping by the bilinear transformation.

magnitude response is transformed into the digital filter using the bilinear transformation.

It should be noted that the bilinear transformation preserves magnitude responses that are piecewise linear and does not preserve the phase response.

6.2.2   Lowpass Digital Filter Design

In practice the specifications of the digital filter are provided in terms of the following parameters:

The passband-edge frequency $= \omega_P$
The stopband-edge frequency $= \omega_s$
The maximum passband attenuation $= R_p$, and
The minimum stopband attenuation $= R_s$.

If we design the analog lowpass filter using the digital frequencies ω_P and ω_s we will obtain a transfer function $H_a(s)$ for the filter. The digital filter can be obtained by direct substitution of s by $\frac{1-z^{-1}}{1+z^{-1}}$ in the analog transfer function $H(s)$ so that $H(z) = H_a(s)|_s = \frac{1-z^{-1}}{1+z^{-1}}$. If we use this process the resulting frequencies of the digital filter will be incorrect because ω_P and ω_s will be warped to other frequencies. For this reason the digital band-edge frequencies must be pre-warped to obtain the analog frequencies using the relation $\Omega = \tan(\omega/2)$. The design procedure using the bilinear transformation, therefore is according to the following steps:

(i) Pre-warp the band-edge frequencies (ω_P and ω_s) to obtain the corresponding analog frequencies (Ω_P and Ω_S) of the prototype lowpass filter. The maximum passband attenuation R_P and the minimum stopband attenuation R_s remain unchanged.

(ii) Using procedures for analog filter design in Chapter 5 compute the order N and the cut-off frequency Ω_c.

(iii) Using procedures for analog filter design in Chapter 5 determine the transfer function $H_a(s)$.

(iv) The transfer functions read from [1] given the filter order have been normalized with respect the cut-off frequency. They must be scaled up in frequency such that $\hat{H}(s) = H(s/\Omega_c)$.

(v) Apply the bilinear transformation to obtain the digital filter transfer function $H(z) = \hat{H}(s)\big|_{s=\frac{1-z^{-1}}{1+z^{-1}}}$.

Example 6.1 (Lowpass Filter Design). Design a digital Butterworth lowpass filter to have the passband edge frequency of 10 Hz, a stopband edge frequency of 90 Hz, a maximum passband attenuation of 0.5 dB and a minimum passband attenuation of 20 dB. The sampling frequency is 200 Hz (see Figure 6.4).

Solution

(i) Compute the normalized digital frequencies

$$\omega_P = 2\pi \frac{10}{200} = 0.314\,\text{rad/s} \quad \text{and}$$

$$\omega_s = 2\pi \frac{60}{200} = 1.8850\,\text{rad/s}$$

Pre-warp the digital frequencies to obtain the frequencies of the analog LPF

$$\Omega_P = \tan(0.314/2) = 0.1583\,\text{rad/s} \quad \text{and}$$

$$\Omega_s = \tan(1.8850/2) = 1.3764\,\text{rad/s}$$

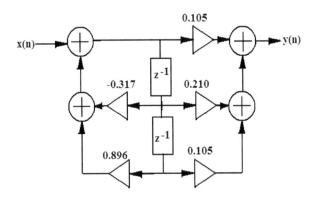

Fig. 6.4 Realization diagram of Example 6.1

(ii) Find the order of the LPF

The inverse transition ratio $\frac{1}{k} = \frac{\Omega_s}{\Omega_P} = \frac{1.3764}{0.1583} = 8.6949$.

From the specified ripple of 0.5 dB we have $10 \log_{10} \left(\frac{1}{1+\varepsilon^2} \right) = -0.5$ which gives $\varepsilon^2 = 0.1220$. From the minimum stopband attenuation we have $10 \log_{10} \frac{1}{A^2} = -20$ which gives $A^2 = 100$. Therefore, the inverse discrimination ratio is given by $\frac{1}{k_1} = \frac{\sqrt{A^2-1}}{\varepsilon} = 28.4864$. The order of the filter is given by

$$N = \frac{\log_{10} \left(\frac{1}{k_1} \right)}{\log_{10} \left(\frac{1}{k} \right)} = \frac{\log_{10} 28.4864}{\log_{10} 8.6949} = 1.55.$$

The order is rounded up to the next higher integer, $N = 2$. In order to determine Ω_c and to meet the stopband specifications exactly we use $\frac{1}{1+(\Omega_P/\Omega_C)^{2N}} = \frac{1}{A^2}$ which gives $\Omega_c = 0.4364$.

(iii) The normalized transfer function of the second-order analog filter is given by $H(s) = \frac{1}{s^2+\sqrt{2}s+1}$. In order to frequency scale the transfer function we replace s with $\frac{s}{\Omega_c}$ which gives

$$\hat{H}(s) = \frac{\Omega_c^2}{s^2 + \Omega_c\sqrt{2}s + \Omega_c^2} = \frac{0.1904}{s^2 + 0.6172s + 0.1904}.$$

(iv) Applying the bilinear transformation we get

$$H(z) = \hat{H}\Big|_{s=\frac{1-z^{-1}}{1+z^{-1}}} = \frac{0.1904(1 + 2z^{-1} + z^{-2})}{1.8076 - 1.6192z^{-1} + 0.5732z^{-2}}$$

$$= \frac{0.105 + 0.210z^{-1} + 0.105z^{-2}}{1 - 0.896z^{-1} + 0.317z^{-2}}.$$

(v) Implementation using direct form II structure

The problem can also be solved using MATLAB. The functions used are shown in Program 6.1 in the appendix. In the bilinear transformation one can cancel out the impact of T by using $T = 2$. The program does the pre-warping outside the bilinear function and therefore the pre-warping option is not used.

Results printed by the program

$\Omega p = 0.1584\,\text{rad/s}$, $\Omega s = 1.3764\,\text{rad/s}$,

Cut-off frequency of the analog filter $= 0.4363\,\text{rad/s}$

Order of the analog filter $= 2$

Transfer function of the analog filter

$$H(s) = \frac{0.1904}{s^2 + 0.6171s + 0.1904}$$

Transfer function of the digital filter

$$H(z) = \frac{0.1053z^2 + 0.2107z + 0.1053}{z^2 - 0.8958z + 0.3172}.$$

6.2.3   Design of Highpass, Bandpass, and Bandstop IIR Digital Filters

The procedure for the design of highpass, bandpass, and bandstop filters using the bilinear transform method is summarized in flow Chart 6.1. An example for each type of filter will be given. It is more efficient to use MATLAB for the complete design. The procedure involves the pre-warping of the digital filter frequencies to corresponding analog frequencies of an equivalent analog filter (highpass, bandpass or bandstop). The equivalent analog filter is then spectrally transformed to a normalized prototype low pass filter. The analog prototype filter is designed using the well known analog filter design techniques and its transfer function $H_{LP}(s)$ is computed. By inverse spectral transformation a transfer function of the equivalent filter $H_D(s)$ is obtained. By using the bilinear transformation, discussed above, $H_D(s)$ is transformed to a digital filter. Table 6.1 gives the various spectral transformations required and flow Charts 6.1 and 6.2 give a summary of the design procedure at a glance for the Highpass and bandpass digital filter.

Example 6.2 (Highpass Filter Design). Design a digital Chebyshev type II highpass filter using the bilinear transform method to suppress the bass component of music coming from a HiFi system. The spectrum of the bass component of music must be attenuated from

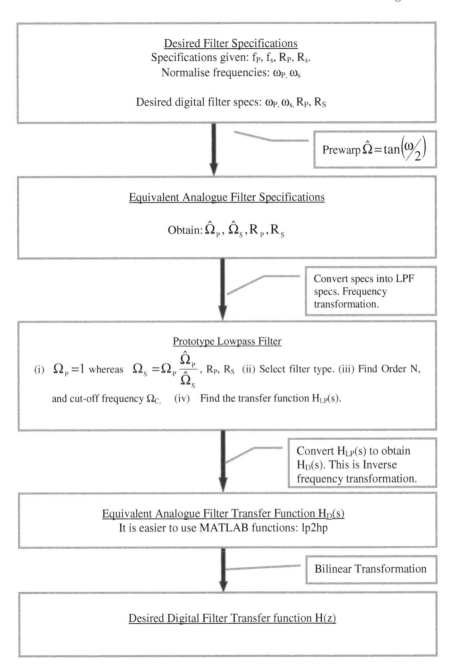

Chart 6.1 Procedure for the design of a digital highpass filters.

Table 6.1 Spectral transformations.

From Lowpass to	Spectral transformation	MATLAB command to implement inverse spectral transformation
Highpass	$\Omega = \Omega_P \dfrac{\hat{\Omega}_P}{\hat{\Omega}}$	$[BT, AT] = \text{lp2hp}[B, A, W_P]$
Bandpass	$\Omega = -\Omega_P \dfrac{\hat{\Omega}_P^2 - \hat{\Omega}^2}{\hat{\Omega} Bw}$	$[BT, AT] = \text{lp2bp}[B, A, W_0, Bw]$
Bandstop	$\Omega = -\Omega_S \dfrac{\hat{\Omega} Bw}{\hat{\Omega}_0^2 - \hat{\Omega}^2}$	$[BT, AT] = \text{lp2bs}[B, A, W_0, Bw]$

DC to 500 Hz by at least 30 dB. The remaining spectrum starting from 550 Hz must not be attenuated by more than 1 dB (see Figure 6.5).

Solution

Stopband edge frequency $\Omega_s = 500\,\text{Hz}$

Passband edge frequency $\Omega_p = 550\,\text{Hz}$

The minimum stopband attenuation $R_s = 30\,\text{dB}$

The maximum passband attenuation $R_P = 1\,\text{dB}$

Select sampling frequency $Fs = 1200\,\text{Hz}$.

Using MATLAB the following results are obtained using Program 6.2 in the Appendix.

Results

Prototype LPF

Cut-off frequency $Wc = 1.8190$, Order $N = 4$

Transfer function:

$$H(s)_{\text{LP}} = \frac{\begin{matrix} 0.03162s^4 - 8.397e - 017s^3 \\ +0.8371s^2 - 9.183e - 016s + 2.77 \end{matrix}}{s^4 + 3.21s^3 + 5.179s^2 + 4.904s + 2.77}.$$

Analog Highpass filter

Transfer function:

$$H(s)_{\text{HP}} = \frac{\begin{matrix} s^4 - 2.528e - 015s^3 + 39.48s^2 \\ -1.118e - 012s + 194.8 \end{matrix}}{s^4 + 20.24s^3 + 244.3s^2 + 1730s + 6161}.$$

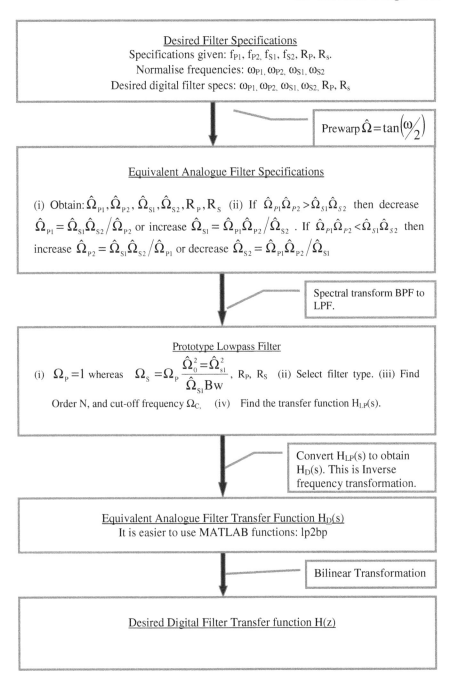

Chart 6.2 Procedure for the design of a digital bandpass filters.

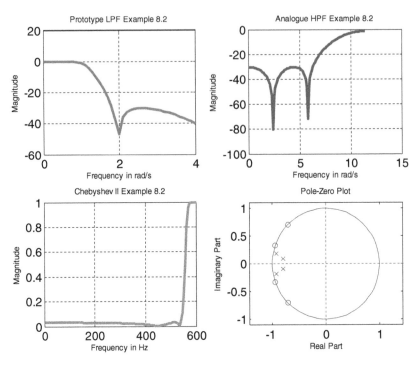

Fig. 6.5 Magnitude response for the digital filter and the intermediate analog filters and the pole–zero plot for the digital filter.

Digital Highpass filter

Transfer function:

$$H(z) = \frac{0.02885z^{\wedge}4 + 0.09505z^{\wedge}3 + 0.1344z^{\wedge}2 + 0.09505z + 0.02885}{z^{\wedge}4 + 3.44z^{\wedge}3 + 4.473z^{\wedge}2 + 2.601z + 0.5707}.$$

Example 6.3 (Bandpass Filter Design). A measuring equipment is picking a telemetry signal using an ultrasonic carrier at 40 KHz. In order to limit the channel noise the ultrasonic signal is filtered using a 10 KHz 1 dB bandwidth filter which is implemented in software using a digital signal processing device. If the maximum attenuation in the passband is not to exceed 0.5 dB and if the minimum attenuation at ±20 kHz from the center of the filter is to be 40 dB design an elliptic IIR digital filter using the bilinear transformation method to perform the task (see Figure 6.6).

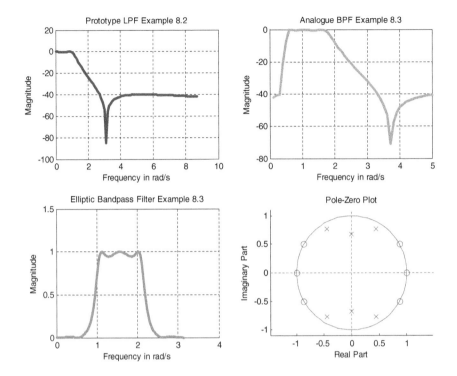

Fig. 6.6 Plots for Example 6.3.

Solution

A bandpass filter is required.

The lower passband edge frequency $f_{P1} = 35\,\text{KHz}$

The upper passband edge frequency $f_{P2} = 45\,\text{KHz}$

The lower stopband edge frequency $f_{s1} = 20\,\text{KHz}$

The upper stopband edge frequency $f_{s2} = 60\,\text{KHz}$

The maximum passband attenuation $R_p = 0.5\,\text{dB}$

The minimum stopband attenuation $R_s = 40\,\text{dB}$

Let the sampling frequency is $120\,\text{KHz}$

Using MATLAB program the following results are obtained using Program 6.3 in the appendix.

Results

Prototype LPF

Cut-off frequency $Wc = 1$; Order $N = 3$

Transfer function:

$$H_{\mathrm{LP}}(s) = \frac{0.07845s^{\wedge}2 - 2.954e - 017s + 0.7555}{s^{\wedge}3 + 1.24s^{\wedge}2 + 1.529s + 0.7555}$$

Analog Bandpass filter
Transfer function:

$$H_{\mathrm{BP}}(s) = \frac{\begin{array}{c}0.08716s^{\wedge}5 - 7.478e - 017s^4 + 1.21s^3 - 6.669e - 017s^2 \\ +0.08716s + 1.007e - 019\end{array}}{s^6 + 1.377s^5 + 4.887s^4 + 3.791s^3 + 4.887s^2 + 1.377s + 1}$$

Digital Bandpass filter
Transfer function:

$$\frac{\begin{array}{c}0.07558z^6 - 7.215e - 016z^5 - 0.1506z^4 + 1.197e - 015z^3 \\ +0.1506z^2 + 1.181e - 015z - 0.07558\end{array}}{\begin{array}{c}z^6 + 3.858e - 015z^5 + 1.235z^4 + 2.644e - 015z^3 \\ +0.973z^2 + 8.552e - 016z + 0.2854\end{array}}$$

The design of a bandstop filter is similar to the design of a bandpass filter with the following difference:

(i) The bandwidth of the equivalent bandstop filter $B_w = \Omega_{S2} - \Omega_{S1}$.

(ii) For the prototype LPF $\Omega_S = 1$ and Ω_P has to be calculated.

(iii) The MATLAB function for inverse frequency transformation in lp2bs.

Example 6.4 (Bandstop Filter Design). A student's project involves transmitting data and speech over a telephone line in the same voice band from a Security Company to a specific subscriber. This data carries information that can be used to display intruded zones. The student is to use a specified band 800 Hz to 1.2 KHz for data transmission and the rest of the band should carry voice in the normal way. The proposed transmission system is shown in Figure 6.7.

Design the bandstop filter for this project to have the following specifications (see Figures 6.8 and 6.9).

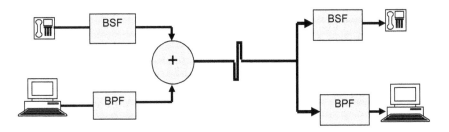

Fig. 6.7 Inband data and voice transmission system over a telephone line.

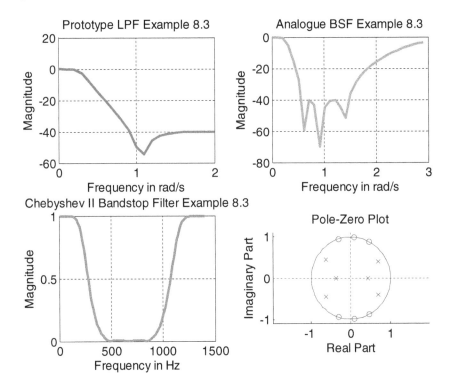

Fig. 6.8 Bandstop filter to for Example 6.4.

Lower passband edge frequency = 750 Hz
Upper passband edge frequency = 1.25 KHz
Lower stopband edge frequency = 800 KHz
Upper stopband edge frequency = 1.2 KHz
Maximum passband attenuation = 0.5 dB.

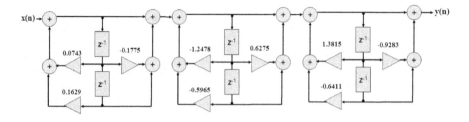

Fig. 6.9 Realization diagram for Example 6.4.

Minimum passband attenuation = 40 dB
Select a sampling frequency = 2.8 KHz.

Solution

Using MATLAB Program 6.4 in the appendix.

Results

Prototype LPF
Cut-off frequency Wc = 0.9149, Order N = 3
Transfer function:

$$H_{LP}(s) = \frac{0.02745s^2 + 0.03063}{s^3 + 0.6172s^2 + 0.1901s + 0.03063}.$$

Analog Bandstop filter
Transfer function:

$$\frac{\begin{array}{c} s^6 + 2.175e - 016s^5 + 3.117s^4 + 1.134e - 016s^3 \\ + 2.609s^2 - 9.191e - 017s + 0.5864 \end{array}}{s^6 + 5.104s^5 + 16.14s^4 + 26.71s^3 + 13.51s^2 + 3.576s + 0.5864}$$

Digital Bandstop filter
Transfer function:

$$\frac{\begin{array}{c} 0.1097z^6 - 0.05249z^5 + 0.2712z^4 - 0.09364z^3 \\ + 0.2712z^2 - 0.05249z + 0.1097 \end{array}}{\begin{array}{c} z^6 - 0.208z^5 - 0.6393z^4 + 0.03381z^3 + 0.4634z^2 \\ - 0.02448z - 0.0623 \end{array}}$$

sos =

$$
\begin{array}{cccccc}
1.0000 & -0.1775 & 1.0000 & 1.0000 & -0.0743 & -0.1629 \\
1.0000 & 0.6275 & 1.0000 & 1.0000 & 1.2478 & 0.5965 \\
1.0000 & -0.9283 & 1.0000 & 1.0000 & -1.3815 & 0.6411
\end{array}
$$

g = 0.1097.

6.3 FIR Filter Design

FIR filter are not designed from analog filters. They are designed by direct approximation of the magnitude or the impulse response and may have a condition that the phase frequency characteristic be linear. There are two common methods that are used; the windowed Fourier series method and the frequency sampling method. In this section, we will focus on the windowed Fourier series method.

6.3.1 The Windowed Fourier Series Method

FIR filters have a transfer function $H(z)$ that is a polynomial of z^{-1} and is given as the z-transform of its impulse response

$$
H(z) = \sum_{n=-\infty}^{\infty} h(n)z^{-n}. \tag{6.6}
$$

When Equation (6.6) is evaluated on the unit circle, $z = e^{j\omega}$, we obtain the frequency response (or the DTFT) of the filter given as

$$
H(e^{j\omega}) = \sum_{n=-\infty}^{\infty} h(n)e^{-j\omega n}. \tag{6.7}
$$

The frequency response of the discrete-time signal is periodic and therefore the expression of Equation (6.7) is a Fourier series and $h(n)$ are the Fourier series coefficients which can be given as

$$
h(n) = \frac{1}{2\pi} \int_{-\pi}^{\pi} H(e^{j\omega})e^{j\omega n} d\omega. \tag{6.8}
$$

Normally we aim at magnitude response of an ideal filter with a "brick wall" response and this has an infinite length impulse response with samples covering the range $-\infty < n < \infty$. Such a filter is noncausal and therefore unrealizable. The filter is also unstable. The design process involves shortening the filter to a desired length and delaying it to make it causal. The process of shortening the filter length is referred to as truncation or windowing and is achieved by multiplication by window function.

Let the truncated impulse response be denoted by $h_t(n)$ covering the range $-N \leq n \leq N$. It is intended to make the DTFT $(h_t(n))$ (or $H_t(e^{j\omega})$) approximate $H(e^{j\omega})$ so that the integral-squared error is minimized. If we apply Parsevals' relation we can write

$$\varepsilon = \frac{1}{2\pi} \int_{-\pi}^{\pi} |H(e^{j\omega}) - H_t(e^{j\omega})|^2 d\omega$$

$$= \sum_{n=-\infty}^{\infty} |h(n) - h_t(n)|^2$$

$$= \sum_{n=-N}^{N} |h(n) - h_t(n)|^2 + \sum_{n=-\infty}^{-N-1} |h(n)|^2 + \sum_{n=N+1}^{\infty} |h(n)|^2, \quad (6.9)$$

which shows that the integral squared error is minimized when $h_t(n) = h(n)$ *in the range* $- N \leq n \leq N$. This implies that the process of truncation results in a minimum integral squared error. In order to make the filter causal we can delay the impulse response by N such that the delayed samples are given by $h_1(n) = h_t(n - N)$. Delaying the samples does not change the magnitude response but change the phase response. The design procedure for FIR filters then involves truncating the infinite-length impulse response $h(n)$ of ideal filters and shifting such impulse responses to make the filter causal. The expressions for the impulse responses of ideal filters is summarized in Table 6.2.

6.3.2   The Gibbs Phenomenon

When an ideal impulse response $h(n)$ of infinite length is truncated by multiplication by a window function $w(n)$ of finite-length the transfer function of the ideal filter which was originally rectangular shows oscillatory behavior. This oscillatory behavior is known as Gibbs phenomenon. In order to explain Gibbs phenomenon the following functions are specified:

Table 6.2 Impulse responses of ideal filters.

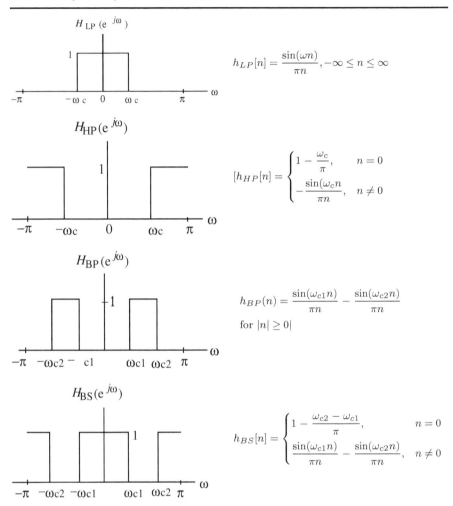

$$h_{LP}[n] = \frac{\sin(\omega n)}{\pi n}, -\infty \leq n \leq \infty$$

$$[h_{HP}[n] = \begin{cases} 1 - \dfrac{\omega_c}{\pi}, & n = 0 \\ -\dfrac{\sin(\omega_c n)}{\pi n}, & n \neq 0 \end{cases}$$

$$h_{BP}(n) = \frac{\sin(\omega_{c1} n)}{\pi n} - \frac{\sin(\omega_{c2} n)}{\pi n}$$

for $|n| \geq 0|$

$$h_{BS}[n] = \begin{cases} 1 - \dfrac{\omega_{c2} - \omega_{c1}}{\pi}, & n = 0 \\ \dfrac{\sin(\omega_{c1} n)}{\pi n} - \dfrac{\sin(\omega_{c2} n)}{\pi n}, & n \neq 0 \end{cases}$$

The impulse response of the ideal filter is denoted by $h(n) - \infty < n < \infty$. The window function, in this case a rectangular window, is defined as

$$w(n) = \begin{cases} 1 & -M \leq n \leq M \\ 0 & \text{otherwise} \end{cases}, \tag{6.10}$$

which has a DTFT given by

$$W(e^{j\omega}) = \frac{\sin\left(\frac{\omega}{2}(2M+1)\right)}{\sin\left(\frac{\omega}{2}\right)}. \tag{6.11}$$

The truncated impulse response will thus be defined as

$$h_t(n) = h(n)w(n) = \begin{cases} h(n) & -M \le n \le M \\ 0 & \text{otherwise} \end{cases} \tag{6.12}$$

Using the modulation property of DTFT it is observed that time domain multiplication implies convolution in the frequency domain. This we can write

$$H_t(e^{j\omega}) = \frac{1}{2\pi}\int_{-\infty}^{\infty} H(e^{j\varphi})W(e^{j(w-\varphi)})d\varphi, \tag{6.13}$$

and is shown graphically in Figure 6.10. The DTFT of the window function resembles a sinc function with a main lobe and smaller side lobes. In the process of convolution a frequency reversal of this function represented by $W(e^{-j\varphi})$ is shifted to the right as $W(e^{j(w-\varphi)})$ by an amount w and the area under the curve $y = W(e^{j(w-\varphi)}) \times H(e^{j\varphi})$ from $-\omega_c$ to ω is taken. As ω increases the area under the curve y will be increasing, and varying according to whether the lobes are shifting in or not. The contribution from the main lobe is a slow transition to a maximum and when fully in the overlapping region its contribution is constant defining the passband of the truncated filter. As the main lobe moves out of the overlapping region its contribution is defined by a slow transition to zero. Meanwhile the sidelobes add ripples on top of the main lobe contribution. The magnitude response of the truncated impulse response has the following characteristics:

(i) There are ripples in both the passband and the stopband.
(ii) If the filter length is increased the number of ripples in both the passband and the stopband increase with a corresponding decrease in the ripple width.
(iii) If a sharp transition is required the main lobe width has to be as small as possible. The main lobe width obtained from the first zero crossings in Equation (6.11) is $4\pi/(2M+1)$. Thus

Multiplication in the time domain	Frequency convolution
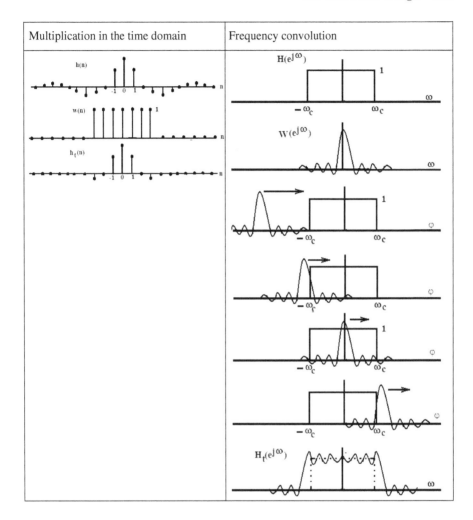	

Fig. 6.10 The windowing process shown graphically in the time and frequency domains.

for a sharp transition M has to be as large as possible. However, this is undesirable as a large window length and hence a large filter length, increases computational complexity.

It is the sharp transition to zero of the window outside $-N \leq n \leq +N$ that causes the Gibbs phenomenon in the magnitude response of the windowed truncated filter. In order to reduce the Gibbs phenomenon windows that tapers smoothly to zero on either side can be used.

Such windows reduce the height of the sidelobes and increase the main lobe width. When the window DTFT is convolved with the magnitude response of the ideal filter the result is a more gradual magnitude response of the truncated filter.

6.3.3   Window Functions

Important parameters in the design of FIR filters using window functions:

(i) Let w_p and w_s be the passband and stopband edge frequencies of the desired filter. Then the cut-off frequency $w_c = \frac{w_p + w_s}{2}$ and $H(e^{jw_c}) = 0.5$.

(ii) The transition bandwidth $\Delta w = w_s - w_p$ and is approximately given by $\Delta w \cong C/M$.

(iii) The transition width of the window is always less than the main lobe width.

In the following section, the window analysis tool "wintool" is used to analyze selected windows. There are several other windows functions in the toolbox.

(i) Rectangular window

$$w(n) = \begin{cases} 1 & -M \leq n \leq M \\ 0 & \text{otherwise.} \end{cases} \tag{6.14}$$

The graphical representation of rectangular window is shown in Figure 6.11.

Main Lobe width $= 4\pi/(2M + 1)$, Relative Sidelobe Level $= 13.3\,\text{dB}$.

Transition bandwidth $= 0.92\,\pi/\text{M}$, Minimum stopband attenuation $= 20.9\,\text{dB}$.

(ii) The Hann window

$$w(n) = \frac{1}{2}\left(1 + \cos\left(\frac{2\pi n}{2M + 1}\right)\right) \quad \text{for } -M \leq n \leq M \tag{6.15}$$

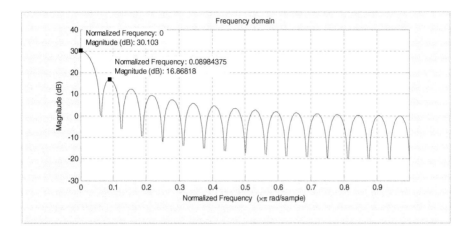

Fig. 6.11 The frequency response of a rectangular window for $M = 32$.

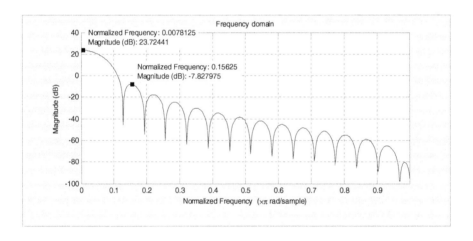

Fig. 6.12 Frequency response of a Hann window for $M = 32$.

The graphical representation of the Hann window is shown in Figure 6.12.

Main Lobe width $= 8\pi/(2M + 1)$, Relative Sidelobe Level $= 31.5$ dB.

Transition bandwidth $= 3.11 \ \pi/M$, Minimum stopband attenuation $= 43.9$ dB.

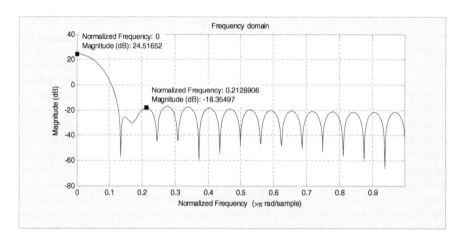

Fig. 6.13 Frequency response of a Hamming window for $M = 32$.

(iii) Hamming Window

$$w(n) = 0.54 + 0.46\cos\left(\frac{2\pi n}{2M + 1}\right) \quad \text{for } -M \le n \le M.$$
$$(6.16)$$

The graphical representation of the Hamming window is shown in Figure 6.13.

Main Lobe width $= 8\pi/(2M + 1)$, Relative Sidelobe Level $= 42.7\,\text{dB}$.

Transition bandwidth $= 3.32\ \pi/\text{M}$, Minimum stopband attenuation $= 54.5\,\text{dB}$.

(iv) Blackman window

$$w(n) = 0.42 + 0.5\cos\left(\frac{2\pi}{2M + 1}\right)$$

$$+ 0.08\cos\left(\frac{2\pi}{2M + 1}\right) \quad \text{for } -M \le n \le M. \quad (6.17)$$

The graphical representation of the Blackman window is shown in Figure 6.14.

Main Lobe width $= 12\pi/(2M + 1)$, Relative Sidelobe Level $= 58.1\,\text{dB}$.

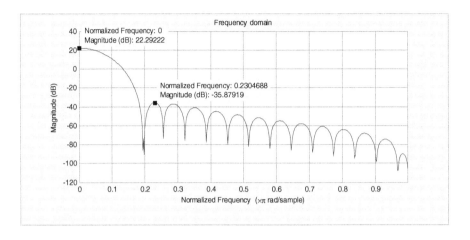

Fig. 6.14 Frequency response of a Blackman window for $M = 32$.

Transition bandwidth $= 5.56\ \pi/\text{M}$, Minimum stopband attenuation $= 73.3\,\text{dB}$.

(v) The Kaiser Window

The main lobe of the Kaiser window hs the most energy for a given relative sidelobe level compared to other windows. This makes it an optimum window as it would provide the smoothest transition from passband to stopband with the smallest ripples. It is also an adjustable window as the window can provide different transition widths for the same window length $L = 2M$. The window function is given by

$$w(n) = \frac{I_0 \lfloor \beta \sqrt{1 - (n - M)^2/M^2} \rfloor}{I_0(\beta)} \quad \text{for } n = 0, 1, \ldots, L - 1$$

(6.18)

where β is an adjustable shape parameter and

$$I_0(\beta) = \sum_{k=0}^{\infty} \left[\frac{\left(\frac{\beta}{2}\right)^k}{k!} \right]^2$$

$=$ zero-order modified function of the first kind.

In this summation only the first 25 terms provide an accurate result.

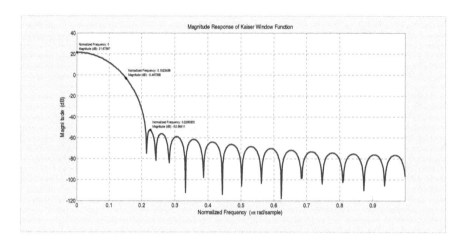

Fig. 6.15 Frequency response of a Kaiser window for $M = 32$ and $\beta = 10$.

The MATLAB function $B = \text{Kaiser}(N, \text{BTA})$ returns N beta valued Kaiser window (see Figure 6.15). The tapering property of the above windows in the time domain is displayed in Figure 6.16. The small difference in the taper produces a significant difference in the relative sidelobe and in the width of the main lobe.

The MATLAB function $b = \text{fir2}(N, f, m, \text{window})$, where N is the order of the filter, f is a vector of band-edge frequency points in the range from 0 to 1 given in increasing order. The variable m is a vector containing the ideal magnitude response at the points specified in f. The vector f and m must be of the same length.

Example 6.5   Design a lowpass FIR digital filter with the following specifications:

The passband edge frequency $f_p = 2\,\text{KHz}$

The stopband edge frequency $f_p = 2.5\,\text{KHz}$

Maximum passband attenuation $= 0.1\,\text{dB}$

Minimum stopband attenuation $= 50\,\text{dB}$

The sampling frequency $F_s = 10\,\text{KHz}$.

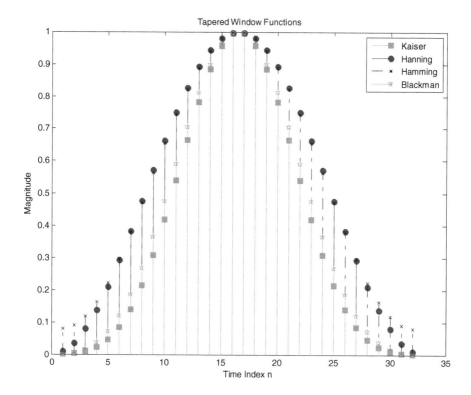

Fig. 6.16 Characteristics of selected tapered window functions.

Solution

$$\omega_p = 2\pi f_p / F_s = 0.5\pi, \quad \omega_s = 2\pi f_s / F_s = 0.4\pi$$

$$\omega_c = \frac{\omega_p + \omega_s}{2} = 0.45\pi, \quad \Delta\omega = \omega_p - \omega_s = 0.1\pi.$$

Select Hamming window since it has $54.5\,\text{dB}$ of stopband attenuation and thus just exceeds the required stopband attenuation. For the Hamming window $\Delta\omega = 3.32\pi/\text{M}$. Hence $M = \frac{3.32\pi}{0.1\pi} = 33.2$. Choose $M = 34$ and therefore $N = 2M + 1 = 69$.

The impulse response of the filter is given by

$$h_t(n) = h_{LP}(n) \times w(n)|_{\text{Hamming}} = \left(\frac{\sin(\omega n)}{\pi n} \right)$$

$$\times \left(0.54 + 0.46 \cos\left(\frac{2\pi n}{2M+1} \right) \right) \quad \text{for } -M \leq n \leq M$$

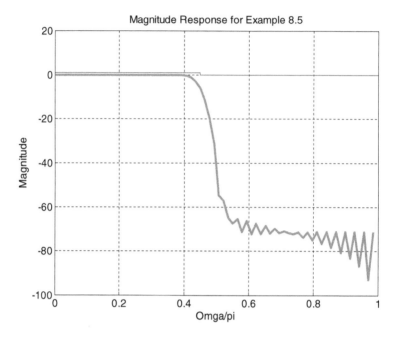

Fig. 6.17 Magnitude response of an FIR filter designed using the Hamming window.

MATLAB Programme to design FIR filter using the Hamming window.

```
% FIR filter Design using the Hamming window
% f specifies the position of the band-edges and
  m the ideal
% magnitude at those positions
f = [0 0.45 0.45 1]; m = [1 1 0 0];
b = fir2(69,f,m, hamming(70));
[h,w] = freqz(b,1,69);
plot(f,m,w/pi,20*log10(abs(h)))
legend('ideal','fir2 Designed')
title('Comparison of Frequency Response Magnitudes')
```

6.4   Problems

6.5.1 Given the transfer function $H(s) = \frac{1}{(1+3s+s^2)}$, find $H(z)$ using the bilinear transformation method.

6.5.2 Determine the transfer function of a second-order lowpass digital Butterworth filter with a cut-off frequency of 4 KHz and a sampling frequency of 40 KHz.

6.5.3 Determine the transfer function of a second-order highpass digital Butterworth filter with a cut-off frequency of 4 KHz and a sampling frequency of 40 KHz.

6.5.4 Design a highpass digital elliptic filter using the bilinear transform method to meet the following specifications:

  passband-edge frequencies: 4.5 KHz

  stopband-edge frequencies: 4 KHz

  Maximum passband attenuation: 0.5 dB

  Minimum stopband attenuation: 40 dB

Obtain the order, the transfer function, and plot the magnitude and phase response.

6.5.5 Design a bandpass digital Butterworth filter using the bilinear transform method to meet the following specifications:

  passband-edge frequencies: 2 kHz and 4 KHz

  stopband-edge frequencies: 1.8 KHz and 4.5 KHz

  Maximum passband attenuation: 0.5 dB

  Minimum stopband attenuation: 40 dB

Obtain the order, the transfer function, and plot the magnitude and phase response of the digital filter.

6.5.6 Design a bandstop digital Chebyshev II filter using the bilinear transform method to meet the following specifications:

  passband-edge frequencies: 500 Hz and 750 Hz

  stopband-edge frequencies: 550 Hz and 650 Hz

  Maximum passband attenuation: 1 dB

  Minimum stopband attenuation: 50 dB

Obtain the order, the transfer function, and plot the magnitude and phase response of the digital filter.

6.5.7 Another method that is deployed in the design of IIR filters is referred to as the impulse invariance method. In this case the impulse response of the digital filter $h(n)$ is the sampled version of the impulse response of the analog filter $h_a(t)$. That is $h(n) = h_a(nT)$. Show that through the impulse invariance method the causal first-order analog transfer function $H_a(s) = \frac{A}{s+\alpha}$ transforms to a causal first-order digital transfer function given by $G(z) = \frac{A}{1-e^{-\alpha T}z^{-1}}$.

6.5.8 Starting with a numerical approximation to the integral $y(t) = \int_t^{t+T} x(t)dt$ using the trapezoidal rule and the Laplace transform of the integral, obtain the mapping from the s-plane to the z-plane representing the bilinear z-transform.

6.5.9 A first derivative is approximated with a backward difference equation as $\frac{dy(t)}{dt}|_{t=nT} \cong \frac{1}{T}[y(n) - y(n-1)]$, where T is the sampling period and $y(n) = y(nT)$.

  (i) Obtain the mapping from the s-plane to the z-plane.

  (ii) Determine whether stable analog filter will be mapped to stable digital filter.

  (iii) Determine the relationship between the analog frequency Ω and the digital frequency ω.

6.5.10 In this problem we would want to compare the design of an IIR digital filter using the two common methods; the impulse invariance method and the bilinear z-transform method. The same prototype filter should be transformed to a highpass filter and then to a digital filter using the two methods. The specifications of the required highpass digital filter: $\Omega_S = 500\,\text{Hz}, \Omega_P = 550\,\text{Hz}, R_s = 30\,\text{dB}, R_P = 1\,\text{dB}$. Use MATLAB functions to design the digital filter using the two methods. Comment on the magnitude response, the phase response, and the complexity of the two filters.

6.5.11 Design a highpass FIR digital filter with the following specifications:

  The passband edge frequency $f_p = 2.5\,\text{KHz}$

  The stopband edge frequency $f_s = 2\,\text{KHz}$

  Maximum passband attenuation $= 0.1\,\text{dB}$

Minimum stopband attenuation $= 50\,\mathrm{dB}$

The sampling frequency $F_s = 10\,\mathrm{KHz}$.

6.5.12 Design a bandpass FIR digital filter with the following specifications:

The passband edge frequencies $f_{p1} = 2\,\mathrm{KHz}$ and $f_{p2} = 4\,\mathrm{KHz}$

The stopband edge frequency $f_s = 1.5\,\mathrm{KHz}$ and $f_{s2} = 4.5\,\mathrm{KHz}$

Maximum passband attenuation $= 0.1\,\mathrm{dB}$

Minimum stopband attenuation $= 50\,\mathrm{dB}$

The sampling frequency $F_s = 10\,\mathrm{KHz}$.

7

Digital Signal Processing Implementation Issues

7.1 Introduction

Numbers in digital signal processors, and in general, in computers are represented in binary form using a string of symbols 1s and 0s which are referred to as binary digits or in short bits A part of the string on the right representing the integer portion is separated from the part on the left that represents the fractional part by binary point. In general the format takes the form:

$$a_{M-1}a_{M-2}a_{M-3}\cdots a_1 a_0 \triangle a_{-1}a_{-2}\cdots a_{-N} \quad \text{where,}$$

$$a_i = 0 \text{ or } 1 \text{ and } \triangle \text{ is the binary point.}$$

The bit a_{M-1} is referred to as the Most Significant Bit (MSB) and the bit a_{-N} is the least significant bit (LSB). The bit string that represents a specific value is referred to as the word and the number of bits in the word forms the wordlength. The wordlength is normally selected as a power of two; 4, 8, 16, 32, etc.

The decimal value represented by the word is given by

$$\text{Value}_{\text{dec}} = \sum_{i=-N}^{M-1} a^i. \tag{7.1}$$

In digital signal processors numbers are represented in either of two forms; using fixed point numbers or using floating point numbers. In the next two sections, we will discuss these two forms and show how arithmetic operations are achieved.

7.2 Fixed Point Number Representation and Arithmetic

Many DSP processors use fixed-point number representation and arithmetic. In this representation numbers can be represented as integers or

169

as fractions. In the multiplication operations using mixed numbers it is not obvious to determine where the binary point is going to be. However, with products of pure fractions or pure integers the binary point is fixed. For this reason fixed point numbers are always represented as fractions. The first bit is normally used to represent the sign of the fraction. This bit is 1 for negative numbers and 0 for positive numbers. There are three ways in which negative numbers can be represented:

(i) Sign and magnitude representation: The binary number $b = s_\Delta a_{-1} a_{-2} a_{-3} \cdots a_{-N}$ has a positive value given by $\sum_{i=1}^{N} a_{-i} 2^{-i}$ for $s = 0$ or a negative value given by $-\sum_{i=1}^{N} a_{-i} 2^{-i}$ for $s = 1$.

(ii) One's complement representation: The binary number $b = s_\Delta a_{-1} a_{-2} a_{-3} \cdots a_{-N}$ has a positive value given by $\sum_{i=1}^{N} a_{-i} 2^{-i}$ for $s = 0$ (as in sign and magnitude representation) and the negative number is obtained by complementing each bit of the positive fraction. The decimal equivalent of the negative number is obtained from $-(1 - 2^{-N}) + \sum_{i=1}^{N} a_{-i} 2^{-i}$, where a_i are the complemented values.

(iii) Two's complement representation: The binary number $b = s_\Delta a_{-1} a_{-2} a_{-3} \cdots a_{-N}$ has a positive value given by $\sum_{i=1}^{N} a_{-i} 2^{-i}$ for $s = 0$ (as in sign and magnitude representation) and the negative number is obtained by complementing each bit of the positive fraction, and adding a 1 to the LSB. The decimal equivalent of the negative number is obtained from $-1 + \sum_{i=1}^{N} a_{-i} 2^{-i}$, where a_i are the complemented values.

It easier to show arithmetic operations using examples

(i) Addition/Subtraction operations using sign and magnitude representations: Positive numbers are added modulo two column by column. If the sum of any column exceeds one a carry forward to the next column is made. If the addition of two positive numbers results in a negative number, as in Example 7.2, then this indicates an overflow. Subtraction is also done on a column by column basis. If the subtrahend (number to subtract) is bigger than minuend (the number to subtract from) then a borrow is made from the next column. This is shown in Example 7.4.

Example 7.1.

$$
\begin{array}{r}
0_\triangle 1\,0\,1\,1 \\
+\ \ 0_\triangle 0\,1\,0\,0 \\
\hline
0_\triangle 1\,1\,1\,1
\end{array}
$$

Example 7.2.

$$
\begin{array}{r}
\text{carry} \\
0_\triangle 1\,0\,1\,1 \\
+\ \ 0_\triangle 1\,0\,0\,1 \\
\hline
1_\triangle 0\,1\,0\,0
\end{array}
$$

Example 7.3.

$$
\begin{array}{r}
0_\triangle 1\,1\,1\,1 \\
-\ \ 0_\triangle 0\,1\,0\,1 \\
\hline
0_\triangle 1\,0\,1\,0
\end{array}
$$

Example 7.4.

$$
\begin{array}{r}
\text{borrow} \\
0_\triangle 1\,0\,1\,0 \\
-\ \ 0_\triangle 0\,1\,0\,1 \\
\hline
0_\triangle 0\,1\,0\,1
\end{array}
$$

The important observation to make here is that the operation of addition and subtraction are different and would require different circuits or algorithms in their implementation. If in Example 7.4 we treated subtraction as an addition of a negative number it would be necessary to convert the negative number using either one's or two's complement representation as in (ii) and (iii) below.

(ii) Addition/Subtraction operations using one's complement representation: First represent the negative number $-0_\Delta 0101$ as $1_\Delta 1010$ using one's complement representation and add in the normal way. The extra bit created on the left of the sign bit resulting from the carry bit is added to the LSB as shown in Example 7.5. In this case subtraction has been treated as an addition operation.

Example 7.5.

$$0_\Delta 1\,0\,1\,0 \quad \text{is equivalent to}$$
$$- \quad 0_\Delta 0\,1\,0\,1$$
$$\overline{0_\Delta 0\,1\,0\,1}$$

$$0_\Delta 1\,0\,1\,0$$
$$+ \quad 1_\Delta 1\,0\,1\,0 \qquad \text{one's complement}$$
$$\overline{1\,0_\Delta 0\,1\,0\,0} \qquad \text{of the sutrahend}$$
$$+ \qquad\qquad 1$$
$$\overline{0_\Delta 0\,1\,0\,1} \quad \text{end around carry}$$

(iii) Addition/Subtraction operations using two's complement representation: In Example 7.6 the same subtraction is implemented as an addition operation of a negative number only this time the negative number is represented using two's complement representation.

Example 7.6.

$$0_\Delta 1\,0\,1\,0 \quad \text{is equivalent to}$$
$$- \quad 0_\Delta 0\,1\,0\,1$$
$$\overline{0_\Delta 0\,1\,0\,1}$$
$$\text{drop the carry}$$

$$0_\Delta 1\,0\,1\,0$$
$$+ \quad 1_\Delta 1\,0\,1\,1 \qquad \text{two's complement}$$
$$\overline{1\,0_\Delta 0\,1\,0\,1} \qquad \text{of the sutrahend}$$
$$0_\Delta 0\,1\,0\,1 \qquad \text{correct answer}$$

The conclusion here is that with negative numbers represented in one's complements form or two's complement form subtraction can be treated as an addition operation and hence the same algorithm or circuit can be used for addition and for subtraction.

7.2.1 Fixed Point Multiplication

Consider two binary numbers to be multiplied $x = x_s \triangle x_{-1} x_{-2} x_{-3} \cdots x_{-n}$ (the multiplicand) and $y = y_s \triangle y_{-1} y_{-2} y_{-3} \cdots y_{-n}$ (the multiplier), where x_s and y_s represent the sign of the respective numbers. The multiplication can be performed in n steps using the following algorithm:

$$P^{(i)} = \left(P^{(i-1)} + y_{-n+i-1} \times x\right) \times 2^{-1} \quad \text{for } i = 1, 2, \ldots, -n, \quad (7.2)$$

where $P^{(0)} = 0$. When $y_{-n+i-1}^{=}0$ the current product is equal to the previous product shifted once to the right and when $y_{-n+i-1} = 1$ then the current product is equal to the sum of the previous product and the multiplicand shifted one place to the right. The process is shown in Example 7.7 where the fractions $x = -\triangle 110$ and $y =\triangle 011$ are multiplied. In this case $x_s = -1$ and $y_s = +1$ and the sign of the product will be negative. The process of multiplication of the magnitudes of the numbers is done in the following process:

Example 7.7.

```
      1 1 0      X
      0 1 1      Y
      ─────
      0 0 0      P⁽⁰⁾
      1 1 0
      ─────
      1 1 0
    0 1 1 0      P⁽¹⁾
      1 1 0
    ─────────
  1 0 0 1 0
  1 0 0 1 0      P⁽²⁾
  0 0 0
  ─────────
  1 0 0 1 0
  0 1 0 0 1 0    P⁽³⁾
```

The sign is negative and the magnitude is $\triangle 0\ 1\ 0\ 0\ 1\ 0$

The answer is $-\triangle 0\ 1\ 0\ 0\ 1\ 0$

7.3   Floating Point Number Representation and Arithmetic

A floating point number α is represented using two parameters; the mantissa m and the exponent e such that $\alpha = m \times 2^e$, where the mantissa is a binary fraction in the range $1/2 \leq m < 1$ and the exponent e is a positive or negative binary integer. The IEEE [11] has defined a standard for the floating point number format. For the 32-bit format the first bit is a sign bit, the next 8 bits are used for the exponent and the remaining 23 bits are used for the mantissa as shown in Figure 7.1. Thus a floating point number using the IEEE format is given by

$$\alpha = (-1)^S \times 2^{e-127} \times M. \tag{7.3}$$

In this equation it should be noted that the exponent is biased and it is within the range $-127 < \text{exponent} < 128$ and the mantissa in this scheme has been defined to be in the range $0 \leq m < 1$.

Addition of floating point numbers

(i) Addition of two numbers with the same exponent is easily accomplished by direct addition of the mantissa as shown in Example 7.8.

Example 7.8.

$$\alpha_1 = (0_\triangle 10101100)2^{101}, \quad \alpha_2 = (0_\triangle 01010001)2^{101}$$

$$\text{sum} = \alpha_1 + \alpha_2 = (0_\triangle 10101100 + 0_\triangle 01010001)2^{101}$$

$$= (0_\triangle 1\,1\,1\,1\,1\,1\,0\,1)2^{101}$$

Fig. 7.1 The IEEE 32-bit floating point format.

(ii) In the addition of numbers that have different exponents the mantissa of the smaller numbers is shifted to the right to make the exponent of the smaller number equal to that of the larger number and then the mantissa of the two numbers are added as in (i) above.

Example 7.9.

$$\alpha_1 = (0_\Delta 10101100)2^{101}, \quad \alpha_2 = (0_\Delta 01010001)2^{110}$$

α_1 has the smaller exponent, shift mantissa to the right one step (equivalent to dividing by 2) and add 1 to the exponent (equivalent to multiplying same number by 2).

$$\alpha_1 = (0_\Delta 01010110)2^{110}$$

$$\text{sum} = \alpha_1 + \alpha_2 = (0_\Delta 01010110 + 0_\Delta 01010001)2^{110}$$

$$= (0_\Delta 10100111)2^{101}$$

7.3.1   Multiplication of Floating Point Numbers

Multiplication of floating point numbers is achieved by multiplying their mantissas and adding the exponents. If the product of the mantissa is below the range $1/2 \leq m < 1$ then the mantissa can be normalized by shifting it to the right and making necessary compensation to the exponent.

Example 7.10. Consider two floating point numbers

$$\alpha_4 = (0_\Delta 110)2^{10} \quad \text{and} \quad \alpha_5 = (0_\Delta 011)2^{01}$$

Product of the mantissa is done as in Example 7.7. The resulting product is given by

$$\text{product} = (0_\Delta 010010)2^{11}$$

It can be shown that the Mantissa of the product is equal to 0.28125 in decimal notation. Since this is less than 0.5 normalization is necessary.

Shift the mantissa to the left and subtract 1 from the exponent. This gives the product as

$$\text{product} = (0_\Delta 100100)2^{10}$$

7.4   Fixed and Floating Point DSP Devices

The selection of a fixed-point or floating-point DSP depends, to a large extent, on whether the computational capabilities using the floating point format is required by the particular application. The degree of accuracy required by the application is a significant factor. A good example is in the processing of audio signals versus video signals. Audio signals require much more accuracy than video signals as the ear is more sensitive to details than the eye. Floating point devices are capable of providing the accuracy that audio signals require as greater accuracy is achievable in floating- point devices. On the other hand, in processing video signals a huge amount of data is involved. In order to process such data fast and in real-time fixed-point devices are the most appropriate.

Accuracy in floating point devices is achieved through the use of 24 bits or more for the mantissa to represent the signal variables and the coefficients. In the majority of fixed point devices 16 bits are used [12]. A good example is the TI DSP C67x^{TM} floating-point DSP processor that uses 24 bits would achieve more accuracy than the C62x^{TM} fixed point processor that uses 16 bits. The exponentiation that is employed in floating-point number representation increases the dynamic range available for applications with large and unpredictable dynamic ranges. In addition the use of longer word-lengths to represent the internal product ensures more accuracy to the end product.

There are three important word-lengths in the internal architectures of digital signal processors. The size of these word-lengths are carefully chosen to ensure that the targeted applications are implemented with the desired accuracy. These word-lengths are listed below and summarised in Table 7.1 for selected TI DSPs.

(i) I/O signal word-lengths: In Table 7.1 a sample of TI DSP shows that this word-lengths for fixed-point DSPs may be

Table 7.1 Word-lengths for TI-DSPs [12].

TI DSP(s)	Format	Word-length		
		Signal I/O	Coefficient	Intermediate Result
C25x	Fixed	16	16	40
C5x^{TM}/C62x^{TM}	Fixed	16	16	40
C64x^{TM}	Fixed	8/16/32	16	40
C3x^{TM}	floating	24(mantissa)	24	32
C67x^{TM}(SP)	floating	24(mantissa)	24	24/53
C67x(DP)	floating	53	53	53

16, 24, or 32. In the C64x one can select any of the three. The mantissa of the floating point DSPs, on the hand, can have 24 for the 32-bit format or 53 for the 64-bit format.

(ii) Coefficients word-length: The word-length is identical to the I/O signal word-length except for the C64x DSP where only the 16-bit option is available.

(iii) Intermediate products word-length: For a single 16-bit (from signal) by 16-bit (from coefficient) multiplication a 32-bit product is normally required. Also for a single 24-bit by 24-bit multiplication a 48-bit product is required. However, the iterated MAC (multiply and accumulate) require additional bits for overflow headroom. The C62x fixed point DSP the headroom is 8 bits. The intermediate product word-length is therefore 40 bits (16 bits signal + 16 bits coefficients + 8 bits overflow). For the C67x floating point DSP only the mantissa data path is considered. In theory the word-length required is 64-bits (24 bits signal + 24 bits for the coefficient +16 bits overflow). However, this is beyond the accuracy required in most applications. Only 48 bits are kept and due to exponentiation the accuracy achieved is much more than for fixed-point DSPs.

7.5 Overflows Resulting from Arithmetic Operations

Fixed point digital signal processors can experience overflow conditions resulting from addition of signal variables at the intermediate stages as the signal goes through the processor. An overflow leads to severe

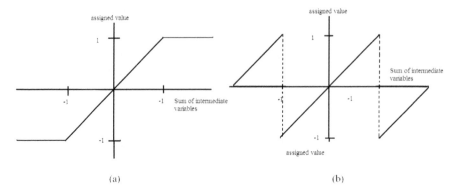

Fig. 7.2 Graphs showing overflow control: (a) saturation overflow; (b) two's complement overflow.

distortion of the output signal and this must be prevented. In order to prevent the distortion when the sum of the signal variables go beyond the expected dynamic range the sum is assigned a different value that is within the range. When the sum exceeds $+1$ it is assigned a value equal to $1 - 2^{-b}$, where b is the number of the LSB and when the sum is below -1 it will be assigned a value of -1. Therefore for the sign and magnitude representation the saturation flow scheme is as shown in Figure 7.2 (a). In two's complement case when the sum goes outside the range then the assigned value is given by

$$\text{assigned value} = \langle \text{sum} + 1 \rangle_2$$

7.6   Impact of the Quantization Process

Analog input to a digital signal processor is converted to a fixed word-length digital signal using an ADC. In Section 7.3, we saw the various word-lengths for the I/O signals for various TI DSP processors. We have also seen that the coefficients are stored in registers with fixed register lengths. Thus the I/O signals are not processed and/or stored in the processor with infinite precision as this would require infinitely large memory. Both the signals and the coefficients are quantized into fixed word-lengths in order to fit into the available storage. During a multiplication operation a b-bit signal sample is multiplied with a b-bit

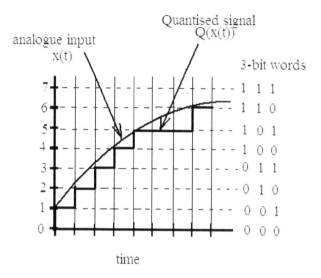

Fig. 7.3 Quantization through truncation.

coefficient to produce a $2b$-bit product. Such a product may be stored or sent to DAC converter after the $2b$ bit word is shortened or truncated to b bits.

The impact of quantization of the input signal, the coefficients and the results of intermediate operations is that the signal at the output will be different from that expected theoretically. This is because quantization introduces errors in the input signal, in the coefficients and in the products of the intermediate operations. The error that is introduced to the output signal depends on how the numbers and the arithmetic operations are done. The errors will be different depending on whether fixed-point or floating-point number representation and arithmetic operations are used. The errors will also be influenced by the way the negative numbers are represented in fixed-point representation.

Figure 7.3 shows a signal $x(t)$ that is quantized through truncation to obtain the signal $Q(x(t))$. The quantization error is given by

$$\varepsilon_t = Q(x) - x. \tag{7.4}$$

In the section below we show the range of the truncation errors for the different type of arithmetic representation.

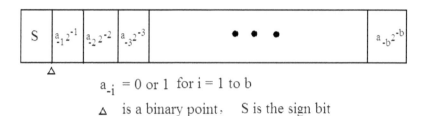

$$a_{-i} = 0 \text{ or } 1 \quad \text{for } i = 1 \text{ to } b$$

$\triangle$ is a binary point.   S is the sign bit

Fig. 7.4 A fixed point fraction represented by b bits.

7.6.1   Quantization Errors in Fixed Point Processors

A fixed-point fraction can be represented as in Figure 7.4 where b bits are used.

(i) Positive numbers: In all the three formats of fixed-point representation the positive numbers are represented in the same way. If, for instance, a positive fraction of length $\rho + 1$ bits is truncated to $b + 1$ bits then the maximum value of the magnitude of the truncation error has a decimal equivalent given by $\varepsilon_t = 2^{-b} - 2^{-\rho}$. This occurs when all bits discarded are ones. The magnitude will be equal to zero if all bits to be discarded are zero. Since the unquantized variable is always greater than the quantized value the error is always negative. Thus the error is bounded by

$$-(2^{-b} - 2^{-\rho}) \leq \varepsilon_t \leq 0. \tag{7.5}$$

For fixed-point negative numbers we have to examine each format used in representing the negative number separately.

(ii) Negative numbers:

- Sign and magnitude representation: For negative fraction the quantized variable is larger than the unquantized variable. Hence the error is positive and is bounded by

$$0 \leq \varepsilon_t \leq 2^{-b} - 2^{-\rho}. \tag{7.6}$$

- For one's complement representation: A negative fraction represented using one's complement representation by the numbers in Figure 7.4 with $S=1$ can have a decimal equivalent value given by $-(1-2^{-b})+\sum_{i=0}^{b}a_i2^{-i}$. This is the value after truncation to $b+1$ bits. If the number of bits before truncation is $\rho+1$ then the numerical value of the negative fraction is $-(1-2^{-\rho})+\sum_{i=0}^{\rho}a_i2^{-i}$. If in the process of truncation only ones are discarded then the maximum quantization error given by the difference between the two is $\varepsilon_t=2^{-b}-2^{-\rho}-\sum_{i=b+1}^{\rho}a_{-i}2^{-i}$. Since we are subtracting a larger negative number from a smaller number the result will always be positive and therefore the quantization error is bounded by

$$0\le\varepsilon_t\le2^{-b}-2^{-\rho}-\sum_{i=b+1}^{\rho}a_{-i}2^{-i}. \qquad (7.7)$$

- For two's complement representation: A negative fraction represented using two's complement representation by the numbers in Figure 7.7 with $S=1$ can have a decimal equivalent value given by $-1+\sum_{i=0}^{b}a_{-i}2^{-i}$. This is the value after truncation to $b+1$ bits. If the number of bits before truncation is $\rho+1$ then the numerical value of the negative fraction is $-1+\sum_{i=0}^{\rho}a_i2^{-i}$. If in the process of truncation only ones are discarded then the maximum quantization error given by the difference between the two is $\varepsilon_t=-\sum_{i=b+1}^{\rho}a_{-i}2^{-i}$, which is always negative. The maximum truncation error is bounded by

$$-(2^{-b}-2^{-\rho})\le\varepsilon_t\le0. \qquad (7.8)$$

(iii) Rounding errors: In rounding the number to be quantized is rounded to the nearest level. Normally only the magnitude is considered and the format in which the negative number is represented is irrelevant. Consider a word that is $1+\rho$ bits long that is to be rounded to the nearest number that is

$1 + b$ bits long, with $\rho > b$. The quantization step is given by $2^{-b} - 2^{-\rho}$. If $|Q(x) - x| \geq \frac{1}{2}(2^{-b} - 2^{-\rho})$, we round up to the next higher magnitude level otherwise we round to the lower magnitude level. It is therefore easy to see that the rounding error range is given by

$$-\frac{1}{2}(2^{-b} - 2^{-\rho}) < \varepsilon_r \leq \frac{1}{2}(2^{-b} - 2^{-\rho}). \qquad (7.9)$$

7.6.2   Quantization Errors in Floating-Point Processors

The format for representing floating-point numbers was given in Equation (7.3). There is no quantization in the exponent as the purpose of the exponent is to increase the dynamic range to a fixed maximum range. Quantization is therefore done only on the mantissa which is a fraction. Equation (7.4) is still valid for the quantization error computation and in terms of the mantissa M we can write

$$\varepsilon = Q(x) - x = 2^E[Q(M) - M]. \qquad (7.10)$$

To determine the range of quantization errors we will consider positive numbers and the different formats in which the negative numbers are represented. Before quantization the mantissa is $\rho + 1$ bits long and after quantization is $b + 1$ bits long. The first bit is the sign bit.

(i) Positive numbers: In all formats the positive numbers are represented in the same way. The truncation error is given by

$$\varepsilon_t = 2^E\left(\sum_{i=1}^{b} a_{-i}2^{-i} - \sum_{i=1}^{\rho} a_{-i}2^{-i}\right) = -2^E\sum_{i=b+1}^{\rho} a_{-i}2^{-i}.$$

Since for positive numbers $Q(M)$ obtained after truncation is always less than the M, ε is always negative. The maximum value of the truncation error is when all the discarded bits are ones and this is given by

$$\varepsilon_t = -2^E(2^{-b} - 2^{-\rho}).$$

Therefore the truncation error range is given by

$$-2^E(2^{-b} - 2^{-\rho}) \leq \varepsilon_t \leq 0. \qquad (7.11)$$

(ii) Negative numbers:

- Sign and magnitude representation: The truncation error is given by

$$\varepsilon_t = -2^E \left(\sum_{i=1}^{b} a_{-i} 2^{-i} - \sum_{i=1}^{\rho} a_{-i} 2^{-i} \right) = 2^E \sum_{i=b+1}^{\rho} a_{-i} 2^{-i}.$$

Since for negative numbers $Q(M)$ obtained after truncation is always larger than the M, ε is always positive. The maximum value of the truncation error is when all the discarded bits are ones and this is given by

$$\varepsilon_t = 2^E (2^{-b} - 2^{-\rho}).$$

Therefore the truncation error range is given by

$$0 \leq \varepsilon_t \leq 2^E (2^{-b} - 2^{-\rho}). \qquad (7.12)$$

- One's complement representation: The truncation error is given by $\varepsilon_t = 2^E \left[(2^{-b} - 2^{-\rho}) - \sum_{i=b+1}^{\rho} a_{-i} 2^{-i} \right]$ and the maximum values is given by $\varepsilon_{t\,max} = 2^E (2^{-b} - 2^{-\rho})$, which is always positive. Therefore the truncation error range is given by

$$0 \leq \varepsilon_t \leq 2^E (2^{-b} - 2^{-\rho}). \qquad (7.13)$$

- Two's complement representation: The truncation error is given by $\varepsilon_t = 2^E \left[-\sum_{i=b+1}^{\rho} a_{-i} 2^{-i} \right]$ and the maximum value of the truncation error is given by $\varepsilon_{t\,max} = -2^E (2^{-b} - 2^{-\rho})$, which is always negative. Therefore the truncation error range is given by

$$-2^E (2^{-b} - 2^{-\rho}) \leq \varepsilon_t \leq 0. \qquad (7.14)$$

(iii) Rounding: Since it is only the mantissa that is subjected to rounding and the range of the rounding error is given by

$$-\frac{1}{2}(2^{-b} - 2^{-\rho})2^E < \varepsilon_r \leq \frac{1}{2}(2^{-b} - 2^{-\rho})2^E. \qquad (7.15)$$

7.6.3   Effects of Coefficient Quantization

There are two significant impacts of quantization on discrete-time systems and in particular digital filters. The transfer function that is obtained after coefficient quantization is different from the theoretical transfer function from the design that assumes infinite length word-lengths. Quantization also has an impact on the poles and zeros position. The location of zeros affects the phase of the system while the location of the poles is more critical as it can make a theoretically stable filter to be practically unstable. A pole that is close to the unit circle in the z-plane can be moved to be on or outside the unit circle making the filter unstable. We will also look at the impact of quantization on the pole positions through simulation.

To quantize coefficients we will use MATLAB that uses the decimal number system. The MATLAB program will have to call functions a2dT and a2dR that will compute the decimal equivalent of the truncated or the rounding numbers. In this investigation we have assumed that the 32-bit long word-lengths for the computer used to run MATLAB represent infinite precision. Figure 7.5(a) shows magnitude response of an elliptical bandpass filter under two conditions. The blue line is the magnitude response when the coefficients are implemented with infinite precision. The red plot is the magnitude response when the coefficients were truncated to 6 bits. It can be observed that after quantization there is more attenuation in the passband and there is a slight shift of the passband to the left. There is also much more attenuation in the stopband. From Figure 7.5(b) we observe that the poles have shifted only slightly but there is a more dramatic move of the zeros. The program to plot the plots of Figure 7.5 are found in the appendix as Program 7.1

7.7   Scaling in Fixed-Point DSPs

Overflows may occur at intermediate stages during signal processing in fixed point discrete-time systems as a consequence of arithmetic operations. The overflow is an output at an intermediate node where addition or multiplication is taking place. Such overflows can cause severe signal

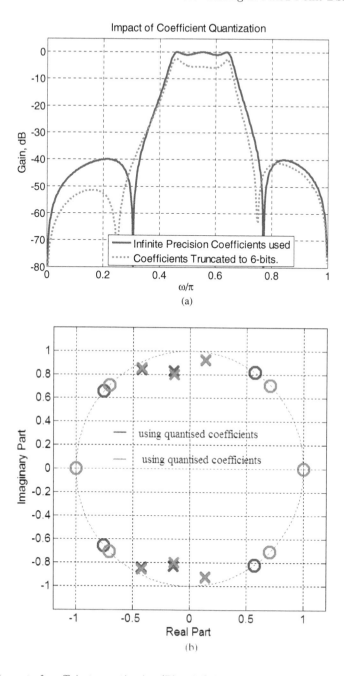

Fig. 7.5 Impact of coefficient quantization (Blue: infinite precision, red: 6 bits).

distortion to the output signal. Chances of the overflow can significantly be reduced if the internal signal levels are scaled to appropriate levels using scaling multipliers inserted at selected positions in the discrete-time system structure. In many cases the scaling multipliers become part of the coefficients and there is not much additional change in the quantization error levels. Consider a discrete-time system whose structure is shown in Figure 7.6.

For the system of Figure 7.6, $x(n)$ is the input signal and $y(n)$ is the output signal from the overall discrete-time system. The output at node i for the same input is given by $w_i(n)$ and the impulse response from the input to the output of node i is given by $g_i(n)$. Thus the output at node i is given by a convolution sum as

$$w_i(n) = g_i(n) \otimes x(n),$$

or

$$w_i(n) = \sum_{k=-\infty}^{\infty} g_i(k)x(n-k). \qquad (7.16)$$

In order to prevent overflows at node i

$$|w_i(n)| \leq 1. \qquad (7.17)$$

We present here three methods and their merits and demerits which can be used to prevent overflows.

(a) Method 1 using the L_1-norm: From Equation (7.16) if we assume that input signal is bounded, i.e., $|x(n)| \leq 1$ it follows

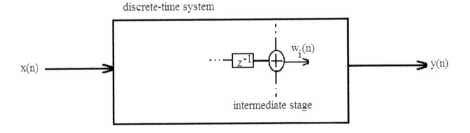

discrete-time system

$x(n)$ $w_i(n)$ $y(n)$

intermediate stage

Fig. 7.6 Discrete-time system showing an intermediate stage.

that

$$|w_i(n)| \leq \left| \sum_{k=-\infty}^{\infty} g_i(k)x(n-k) \right| \leq \sum_{k=-\infty}^{\infty} |g_i(k)| \quad \text{for all } i.$$

(7.18)

If the L_1-norm $= \|g_1\| = \sum_{k=-\infty}^{\infty} |g_i(k)| \leq 1$ then Equation (7.17) is satisfied. A sufficient and necessary condition to guarantee no overflow is that the sum of the absolute values of the impulse response is less than or equal to one. If this condition is not satisfied then the input signal is scaled by multiplying it by a factor K given by

$$K = \frac{1}{\max_i \sum_{k=-\infty}^{\infty} |g_i(k)|}.$$

(7.19)

This is a worst case condition and does not utilize fully the dynamic range of the registers at the adder output. It reduces the signal to noise ratio significantly.

(b) Method 2 using the L_∞-norm: In the frequency domain Equation (7.16) can be written as

$$W_i(e^{j\omega}) = G_i(e^{j\omega})X(e^{j\omega}).$$

(7.20)

From Equation (7.16) we can obtain the $w_i(n)$ by taking the inverse Fourier transform

$$w_i(n) = \frac{1}{2\pi} \int_{-\pi}^{\pi} G_i(e^{j\omega})X(e^{j\omega})e^{j\omega n} d\omega.$$

(7.21)

$$|w_i(n)| \leq \frac{1}{2\pi} \int_{-\pi}^{\pi} |G_i(e^{j\omega})| \|X(e^{j\omega})| d\omega$$

$$|w_i(n)| \leq \frac{\max}{-\pi \leq \omega \leq \pi} |G_i(e^{j\omega})| \frac{1}{2\pi} \int_{-\pi}^{\pi} |X(e^{j\omega})| d\omega$$

$$|w_i(n)| \leq \|G_i(e^{j\omega})\|_\infty \frac{1}{2\pi} \int_{-\pi}^{\pi} |X(e^{j\omega})| d\omega$$

$$|w_i(n)| \leq \|G_i(e^{j\omega})\|_\infty \times \|X\|_1.$$

(7.22)

If $\|X\|_1 \leq 1$, i.e., the mean of the absolute value of the input spectrum is less than one then Equation (7.17) will be satisfied if L_∞-norm $= \|G_i(e^{j\omega})\|_\infty \leq 1$ and there will be no overflow. If this condition is not satisfied then the input signal is scaled by multiplying it by a factor K given by

$$K = \frac{1}{\max|G_i(e^{j\omega})|} \quad \text{for} \quad -\pi \leq \omega \leq \pi. \tag{7.23}$$

The difficulty with this scaling scheme is that it is difficult to obtain an input signal whose magnitude of the spectrum is always less than 1.

(c) Method 3 using the L_2-norm: The magnitude of $w_i(n)$ is bounded by

$$|w_i(n)| \leq \frac{1}{2\pi} \int_{-\pi}^{\pi} |G_i(e^{j\omega})||X(e^{j\omega})|d\omega. \tag{7.24}$$

We can write

$$|w_i(n)|^2 \leq \left(\frac{1}{2\pi} \int_{-\pi}^{\pi} |G_i(e^{j\omega})||X(e^{j\omega})|d\omega \right)^2.$$

Using Schwartz Inequality we obtain

$$|w_i(n)|^2 \leq \left(\frac{1}{2\pi} \int_{-\pi}^{\pi} |G_i(e^{j\omega})|^2 d\omega \right) \times \left(\frac{1}{2\pi} \int_{-\pi}^{\pi} |X(e^{j\omega})|^2 d\omega \right)$$

$$|w_i(n)| \leq \|G_i\|_2 \|X\|_2 \tag{7.25}$$

In Equation (7.25) $\|G_i\|_2 = L_2$-norm for G_i, and $\|X_i\|_2 = L_2$-norm for X representing the input energy of the discrete-time signal. If the input energy has a finite energy bounded by $\|X\|_2 \leq 1$ then the node overflow can be prevented by scaling the discrete-time system such that the rms value of the transfer function from the input to node i is bounded by unity. The input signal and each of the numerator coefficients are multiplied by the scaling factor K given by

$$K = \frac{1}{\text{rms}|G_i(e^{j\omega})|} \quad \text{for} \quad -\pi \leq \omega \leq \pi. \tag{7.26}$$

The results are summarized in Table 7.2.

Table 7.2 Scaling factors for discrete-time systems.

	Norm	Scaling factor K	Location
L_1-norm	$\|g\|_1$	$K = \dfrac{1}{\max_i \sum_{k=-\infty}^{\infty} \|g_i(k)\|}$	From input to k node
L_2-norm	$\|g\|_2$	$K = \dfrac{1}{\left(\sum_{m=0}^{\infty} \|g_k^2(m)\|\right)^{\frac{1}{2}}}$	From input to k node
L_∞-norm	$\|G\|_\infty$	$K = \dfrac{1}{\max \|G_i(e^{jw})\|}$	$-\pi \le w \le \pi$

(d) Scaling using MATLAB simulation: Scaling can be achieved through simulation of the digital filter structure. The simulation is done sequentially following the branch order from the input to the output. Here we will use the L_2-norm. It is easier to use realization structure of the first order and second order. If the impulse response from the input to node i is given by $g_i(n)$ we can implement the scaling according to the following procedure:

(i) Compute the L_2-norm of $g_1(n)$ given by $k_1 = \|g_1\|_2$.

(ii) Divide the input by k_1.

(iii) Compute the L_2-norm of $g_2(n)$ given by $k_2 = \|g_2\|_2$.

(iv) Divide the multiplier feeding to the second adder by k_2.

(v) Continue with the process until the output node gives an L_2-norm $=1$.

A MATLAB program is developed to be used by a cascade of two second-order structures. To obtain a first-order structure the appropriate coefficients are made equal to zero. A cascade of two second-order structures shown in Figure 7.7 is to be scaled.

The transfer function of the system in Figure 7.7 is given by

$$H(z) = \frac{b_{00} + b_{01}z^{-1} + b_{02}z^{-2}}{1 + a_{01}z^{-1} + a_{02}z^{-2}} \times \frac{b_{10} + b_{11}z^{-1} + b_{12}z^{-2}}{1 + a_{11}z^{-1} + a_{12}z^{-2}}. \qquad (7.27)$$

In order to achieve scaling a MATLAB program 7.2 [10] listed in the appendix is used. The program can easily be modified to match any available structure.

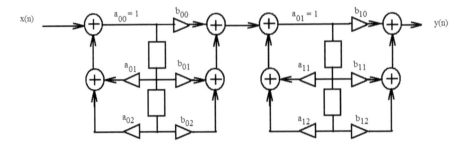

Fig. 7.7 A cascade of two second-order structures.

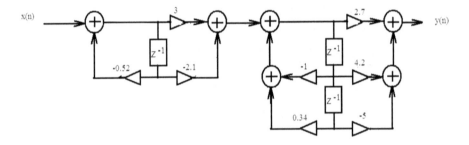

Fig. 7.8 The digital filter structure for the scaling example.

Example 7.11. Using MATLAB simulation scale the digital filter structure is shown in Figure 7.8.

The input coefficients to the program 7.2 in the appendix are shown in Table 7.3. The program computes the scaling factors using the L_2 norm. The scaling factors obtained and the filter structure, with the new coefficients, is shown in Figure 7.9.

Table 7.3 The filter coefficients.

Coefficients for first section		Coefficients for second section	
$a_{00} = 1$	$b_{00} = 3$	$a_{10} = 1$	$b_{10} = 2.7$
$a_{01} = -0.52$	$b_{01} = -2.1$	$a_{11} = -1$	$b_{11} = 4.2$
$a_{00} = 0$	$b_{02} = 0$	$a_{12} = 0.34$	$b_{12} = -5$

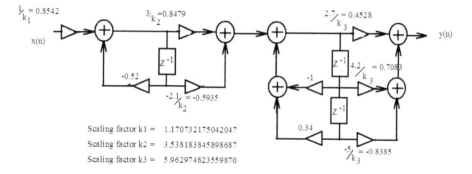

Fig. 7.9 The scaled digital filter.

7.8 Problems

7.7.1 Represent the following decimal fraction in binary form using sign and magnitude representation, one's complement representation, and two's complement representation

(i) 0.65625,   (ii) −0.65625,   and (iii) −0.0500.

7.7.2 Perform the following operation represented by fractions using binary numbers. Subtraction should be performed as an addition of a negative number. Verify the result by comparing to a normal fraction operation.

(i) $\frac{7}{8} - \frac{1}{4}$,   (ii) $\frac{1}{5} - \frac{3}{5}$,   and   (iii) $\frac{3}{5} - \frac{1}{10} - \frac{1}{8}$.

7.7.3 Using the multiplication algorithm for fixed-point binary numbers perform the following multiplication showing all the intermediate stages

(i) $-\frac{3}{8} \times \frac{5}{8}$,   (ii) $\left(-\frac{5}{16}\right) \times \left(-\frac{2}{3}\right)$,   and   (iii) $\left(\frac{5}{16}\right) \times \left(\frac{2}{3}\right)$.

7.7.4 Determine the maximum possible dynamic range for a signal variable when a microprocessor uses floating-point binary numbers with the IEEE format.

7.7.5 Add the following floating-point numbers:

(i) $(0_\Delta 11010101)2^{001}$ and $(0_\Delta 01101010)2^{001}$.
(ii) $(0_\Delta 11010101)2^{011}$ and $(0_\Delta 11010101)2^{001}$.

7.7.6 The maximum possible error when a negative 40-bit binary number obtained after binary multiplication in the ALU is truncated to 16 bits before being stored in the output registers is to be determined. Compute the error if the numbers are represented and the arithmetic operation is done using

(i) sign and magnitude, (ii) one's complement representation or (iii) two's complement representation.

7.7.7 Determine the maximum error that is obtained when the mantissa of a floating point number at an intermediate stage of processing is truncated from 32 bits to 24 bits. Compare all the possible maximum errors for all types of number representations. The truncated number can be a negative number.

7.7.8 A filter has got a transfer function given by $H(z) = \frac{0.44+0.36z^{-1}+0.02z^{-2}}{1+0.4z^{-1}+0.18z^{-2}-0.2z^{-3}}$. Plot the Magnitude responses for infinite precision coefficients and when the coefficients are quantized to 4 bits by truncation. For each case plot the pole positions and comment on the impact of coefficient truncation.

7.7.9 The transfer function $H(z) = \frac{0.44+0.36z^{-1}+0.02z^{-2}}{1+0.4z^{-1}+0.18z^{-2}-0.2z^{-3}}$ is to be implemented using a cascade of a first and second-order transfer function. Determine the following

(i) the new canonic realization structure,

(ii) the scaled coefficients to prevent overflow.

8

Digital Signal Processing Hardware and Software

8.1 Introduction

In this chapter, the DSP hardware and software are introduced. DSP technology is developed by many different groups from academic institutions and industries. There are many common features in DSP hardware and software developed by different vendors. Unfortunately there are also significant differences. The objective of this chapter is to highlight on some common features that can be found in any DSP processor. Since we cannot cover the different features of all the DSP hardware and software from all vendors, we have focussed on hardware and software of a single vendor that we have found in many academic institutions and industry. With a thorough knowledge of the processor from one vendor we believe it will not be so difficult to convert to that of another vendor.

8.2 The Dawn of DSP Processors

Digital signal processing started in the early 1950s out of a different goal; designers of analog systems wished to simulate their designs and investigate their performance before building expensive prototypes. They did this using Digital Computers and this was the beginning of Digital Signal Processing [13]. The majority of the supporting mathematics and algorithms was developed around that time too. Initially, it was sufficient for the simulations to be performed in good time but later some applications required the results in real-time. This ushered in the advent of real-time DSP.

At the moment DSP processors are available as single-chip processors. The evolution of the architecture and technology that led

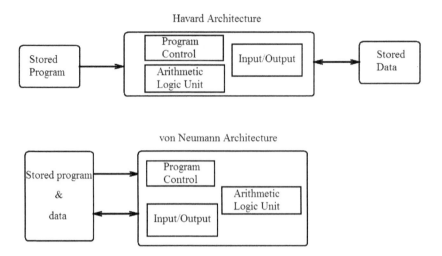

Fig. 8.1 Architectures for processors.

to a single-chip processor took separate paths. Early computers had separate memory spaces for program and data making it possible for each to be accessed simultaneously. This architecture was developed by Howard Aiken at Harvard University and was referred to as the Harvard Architecture, [14], Figure 8.1. This architecture was found to be complex as it had two separate memory spaces. Later a Hungarian mathematician, John von Neumann, observed that there is no fundamental difference between instructions and data. In fact instructions could be partitioned into two fields containing operations commands and addresses of data to be operated upon. Thus only a single memory space was essential for both instructions and data. This architecture was referred to as the von Neumann architecture and processors could be designed with only one memory space [15], Figure 8.1. The drawback of this architecture is that access could be made to either instructions or data at any one time.

The building blocks of a processor include the Arithmetic Logic Unit (ALU), shift registers, and memory space. With these building blocks it is possible to add and subtract by shifting left or right using a few clock cycles but multiply and divide operations are more complex and require a series of shift, add, and subtract operations. Multiply and divide can be achieved in several clock cycles. In the decade starting from early 1970 the multiply times was reduced from 600 ns to 200 ns making real-time digital signal processing a reality [16].

In the early 1970s there was a great need to increase speed, reduce size, and improve processing technologies of electronic devices that were entering the market. This meant increasing the performance and capabilities of integrated circuits. N-MOS was the core integrated circuit technology at the time. With N-MOS it was possible to support device densities of up to 100 000 transistors. With the availability of this technology single chip processors emerged into the market in the early 1980s. With the use of CMOS technology it has been possible to produce single chip processors having up to 4 000 000 transistors with multiply times of 40 ns for 32-bit floating point devices and 25 ns for some 16-bit fixed point devices [13].

8.3   Memory Architectures

It would require to make four accesses to memory in order to implement a simple multiply-accumulate (MAC) instruction, using the von Neumann architecture. The four accesses are made to fetch the instruction, read the signal value, read the coefficient, and write the result. Since the MAC instruction needs to be implemented in one instruction cycle the von Neumann architecture is not suitable for DSP applications in its basic form.

With the Harvard architecture two memory accesses can be made in one instruction cycle. If the architecture is modified such that one memory space holds program instructions and data and the other holds data only then we obtain the "modified" Harvard architecture. Thus the four accesses required in the MAC instruction operation can now be achieved with two instruction cycles. The architecture is shown in Figure 8.2 and is used in some DSP processor families such as Analog Devices ADSP21xx [17].

It is also possible to modify the Harvard architecture further and use three memory spaces each with its own set of buses where one space will be used for program and two for data. The processor can then make three independent memory accesses; one to fetch the MAC instruction

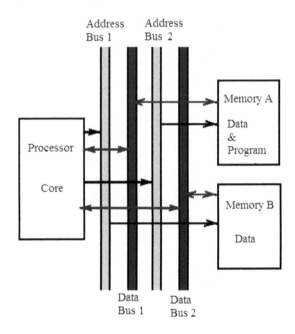

Fig. 8.2 The modified Harvard architecture.

from the program memory, one to read the coefficients and one to read the signal value. The final write instructions can be eliminated by using a different technique referred as modulo addressing. An example of a processor that uses this architecture is Motorola DSP5600x [18].

Yet another alternative to the one proposed above is to use fast memories that support multiple, sequential accesses per instruction cycle over a single set of buses. There are processors with on-chip memories that can complete an access in one half of an instruction cycle. Such processors when combined with the Harvard architecture will be able to make four accesses in one instruction cycle [18].

8.4   Advantages of DSP Processors

There are processors that are made for DSP applications and there other processors that are made for computers or microcontroller and other applications. DSP processors, unlike other processors, are designed to be able to process signals in real time. In order to achieve this there are certain features that are unique to DSP processors only. For instance DSP processors have the ability to multiply and accumulate (MAC) in one instruction cycle. This is achieved by embedding the MAC instruction in hardware in the main data path. Other processors take several instruction cycles to achieve the same operation.

Another feature is that DSP processors have the ability to complete several accesses to memory in a single instruction cycle. For instance, a processor can fetch an instruction while simultaneously storing results of the previous instruction.

Some DSP processors provide special support for repetitive computations, which are typical in DSP computations. A special loop or repeat instruction is provided. Such features make DSP processors more suitable for real-time digital signal processing even when faster processors are available in the market.

Some DSP processors have dedicated address generation units which work in the background and allow the arithmetic processing to proceed with maximum speed. Once the address register is configured it will generate the address required for accessing the operand in parallel with the execution of the arithmetic instruction.

Most DSP processors have one or more serial or parallel input and output (I/O) interface and specialized I/O handling mechanisms such as the direct memory access (DMA). The purpose of these peripherals and interface is to allow a cost effective high performance input and output [7].

8.5 Selection of DSP Processors

Over the years many Digital Signal Processing applications have evolved. Unfortunately there is no single processor that can meet the requirement of each application alone. Also in the market there are many processors manufactured by different vendors. The problem that faces designers is how to select appropriate processors for their specific application. In the discussion below we raise a few issues that will assist a designer in selecting a processor for a specific application.

Many DSP applications today involve portable devices like cellular telephones and portable audio devices. Such devices are battery powered and therefore consumption of power is critical. Suitable processors for such applications would be those processors that consume very little power.

On the other hand, there are applications that involve processing of large volumes of data using complex algorithms such as in seismic detection and sonar. Such applications would require high performance processors in relation to speed, memory requirement, and power consumption. It should be noted that the production volumes for processors for such applications are lower and the cost per processor is high.

The type of arithmetic used in the processor is an important factor in the decision on what processor to use for an application. There are two types of arithmetic format that can be used; fixed point arithmetic and floating point arithmetic. Many DSPs use fixed-point arithmetic, where numbers, n, are represented as fractions in a fixed range; $-1.0 \leq n \leq 1.0$. There are also many processors that use floating-point arithmetic, where values are represented by a mantissa and an exponent as $m \times 2^l$, where m is the mantissa and l is the exponent. The mantissa is generally a fraction in the range -1.0 to $+1.0$, while the exponent is an integer that represents the number of places that the binary point must be shifted left or right in order to obtain the value represented.

Processors using fixed-point arithmetic are normally low power and faster because of the simplicity of implementing fixed-point arithmetic. However, programming of such processors is more complex because of the additional requirement to scale the intermediate signal variables to prevent overflow. On the other hand, processors using floating-point arithmetic have a wider dynamic range for the signal variables. They are easier to program, they consume more power and are relatively slower compared to their fixed-point arithmetic counter part. In the market floating-point DSP processors cost more compared to the fixed-point processors because the more complex floating-point arithmetic require a larger processor, more memory, and more power.

From the above discussion it becomes clear then for high-volume, embedded applications fixed-point processors would be the choice because the focus is on low cost and low power. Applications that require high dynamic range and high precision or where ease of development is significant then floating-point processors are more attractive.

It should also be noted that the word-length to be used has an impact on the cost of the final product. The word-length to be used determines the size of the processor and the number of pins required. It also determines the size of the memory peripheral devices that are connected to the processor. Designers would therefore select the shortest possible word-length if it gives satisfactory performance.

Different applications require different processing speeds. A designer, therefore, needs to determine the processing speed adequate for an application before selecting the processor. The processing speed is measured in terms of the number of million instructions that a processor can perform in one second (MIPS), [19].

8.6 TI DSP Family Overview

There are many DSP vendors in the business and the most established appear to be Analog Devices, Motorola and Texas Instruments (TI) in random order. There are many similarities in the architectures of their processors and the choice of which processor to use and from which vendor is left to the user and his application. In Section 8.4, we have provided a general guideline on their selection which can only be used with the detail data sheet provided for each device. Within each vendor family there are family members that need to be selected for their specific capability. In the next paragraph we will look at the TI TMS320^{TM}DSP family as these are more broadly used than the others.

The TI TMS320^{TM}DSP family can be found under three platforms; TMS320C2000^{TM}DSP platform, TMS320C5000^{TM}DSP platform, and TMS320C6000^{TM}DSP platform. Within each platform there are sub-families to support specific markets. Table 8.1 gives a summary of the general characteristic of each platform and the target market. C2000^{TM}DSP platform is optimized for Digital Control. C5000^{TM}DSP platform are power efficient fixed-point DSP processors finding application in low power portable devices. C6000^{TM}DSP platform represents high performance DSPs operating at high speeds handling more advanced systems involving more complex algorithms.

Table 8.1 The TI TMS320^{TM}DSP family [20].

Platform	Sub-family	Application
C2000	(*high precision*	Control systems
(*Lowest cost*)	*control*)	Digital motor control
	C242, F241,	Digital control
	F242, F243	Digital power supplies
	C2401A, C2402A	Intelligent sensor applications
	C2403A, C2406A	
	C2407A	
	(*fixed point*)	
	F2810, F2812	
C5000	(*fixed point*)	Digital cellular communications
(*Efficient: Best*	C5401, C5402,	Personal communications systems
MIPS/Watt/size/	C5403, C5407,	Pagers
dollar)	C5409, C5410,	Personal digital assistants
	C5416	Digital cordless communications
		Wireless data communications
	C5501, C5502,	Networking
	C5509, C5510	Computer telephony
		Voice over packet
		Portable Internet audio
		Modems
		Digital audio players
		Digital still cameras
		Electronic books
		Voice recognition
		GPS receivers
		Fingerprint/pattern recognition
		Headsets
C6000	(*fixed point*)	Multi-channel (e.g., OFDM)
(*High*	C6201, C6202,	Multi-function applications
performance)	C6203, C6204,	Communication infrastucture
	C6205, C62x^{TM}	Wireless base station
	C6411, C6414,	DSL
	C6415,C6415,	Imaging
	C64x^{TM}	Multimedia services
	(*floating point*)	Video
	C6701,C6711,	
	C6712, C6713,	
	C67x^{TM}	
	C64x^{TM} (*DaVinci*	
	technology)	Digital video systems

8.7   TMS320^{TM}C5416 DSP Processor Architecture [21]

The TMS320^{TM}C5416 processor is a fixed-point digital signal processor that uses an advanced modified Harvard architecture. Data and program are stored in separate memory spaces but by changing the OVLY status certain memory spaces can be set for both program instructions and data. In order to maximize processing power there is one program memory bus against three data memory buses. This makes it possible to have several accesses to memory space in the same cycle. The processor is able to achieve a high degree of processing efficiency because of the following features (Figure 8.3):

(i) Arithmetic Logic Unit (ALU): ALU has a high degree of parallelism. It has a 40-Bit ALU including a 40-bit barrel shifter and two independent 40-bit accumulators (ALU is not shown in the diagram).

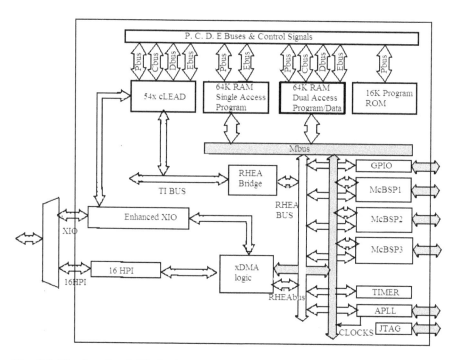

Fig. 8.3 The functional Block diagram of the TMS320CV5416 processor [21]. (*with permission from TI*)

(ii) MAC operation: It has a 17 by 17-Bit (16-bit signed) parallel multiplier coupled to a 40-bit dedicated adder for non-pipelined single-cycle MAC operation.

(iii) Application specific hardware logic: There are several application specific hardware logic such as GSM codec, μ-law compression, a-law compression, and Viterbi accelerator.

(iv) On chip memory: SARAM, DARAM, ROM. The on-chip memory increases performance because it removes the need for wait states and flow within the Central Arithmetic Logic Unit and also this is a lower cost option compared to the use of external memory.

(v) On chip peripherals: Example of these include software programmable wait state generator, a programmable bank switch, a host-port interface (HP18/16), three multichannel buffered serial ports (McBSP), a hardware timer, a clock generator with multiple PLLs, Enhanced extended parallel interface(XIO2), and a DMA controller.

The C5416 processor uses a highly specialized instruction set which allows it to be flexible and accounts for its high speed of operation. Separate program and data spaces allow simultaneous access to program instructions and data, providing the high degree of parallelism. Two reads and one write operation can be performed in a single cycle. Instructions with parallel store and application-specific instructions can fully utilize this architecture. In addition, data can be transferred between data and program spaces. Such parallelism supports a powerful set of arithmetic, logic, and bit-manipulation operations that can all be performed in a single machine cycle.

The C5416 processor has both on-chip RAM and ROM memories which are divided as follows:

(i) Data memory: Addresses up to 64 k of 16-bit words. While within its bounds it accesses on-chip RAM otherwise it will automatically access external memory.

(ii) Program memory: When access is within program address bounds the memory can be accessed directly. If address generated is outside the bounds then an external address is generated. It is possible to configure memory allocation using software to reside inside or outside program address map. By setting the variable OVLY $= 1$ the DARAM memory space can be shared for data and program instructions.

(iii) Extended Program Memory: C5416 uses a paged extended memory scheme in program space to allow up to 8192 K of program memory.

(iv) On-Chip ROM: The on chip ROM consists of 16 K word $\times$ 16-bit maskable ROM that can only be mapped into program. For specific applications request to program the ROM can be made to supplier. A bootloader is available in standard on Chip-ROM and this can be programed to transfer code from an external source to anywhere in the program at power up.

The C5416 has 64 K-word $\times$ 16-bit of on-chip dual access RAM (DARAM) and 64 K-word $\times$ 16-bit of on-chip single-access RAM (SARAM). The DARAM is composed of eight blocks of 8 K words each. Each block in the DARAM can support two reads in one cycle, or a read and a write in one cycle. Four blocks of DARAM are located in the address range 0080h--7FFFh in data space, and can be mapped into program/data space by setting the OVLY bit to one in the PMST register. The other four blocks of DARAM are located in the address range 8000h--FFFFh in program space. The DARAM located in the address range 8000h--FFFFh in program space can be mapped into data space by setting the DROM bit in the PMST register to one.

The SARAM is composed of eight blocks of 8 K words each. Each of these eight blocks is a single-access memory. For example, an instruction word can be fetched from one SARAM block in the same cycle as a data word is written to another SARAM block. This architecture makes possible eight accesses to the SARAM in a single instruction cycle. See Figures 8.3 and 8.4. The SARAM is located in the address range 28000h--2FFFFh, and 38000h--3FFFFh in program space.

Program Space Memory Map

Address (Hex)	Size (Dec)	Function
0x0000	128	Reserved (OVLY=1) External (OVLY=0)
0x007F		
0x0080	32640	On-Chip DARAM 0-3 (OVLY=1) External (OVLY=0)
0x7FFF		
0x8000	32640	External
0xFF7F		
0xFF80	128	Interrupts (On -Chip)
0xFFFF		

Data Space Memory Map

Address (Hex)	Size (Dec)	Function
0x0000	96	Memory mapped registers
0x005F		
0x0060	32	Scratch-Pad RAM
0x007F		
0x0080	32640	On-Chip DARAM 0-3 (32kx16-bit)
0x7FFF		
0x8000	32768	On-chip DARAM 4-7 (DROM=1) or External (DROM=0)
0xFFFF		

Fig. 8.4 Program and data space memory map (Microprocessor mode).

Table 8.2 Memory mapping using the PMST register.

	1	0
MP/MC	On-Chip ROM not available	On-Chip ROM is enabled and addressable
OVLY	On-chip RAM is mapped into program space and data space	On-Chip RAM is addressable in data space but not in program space
DROM	On-Chip DARAM4–7 is mapped into data space	On-Chip DARAM4–7 is not mapped into data space

The three bits of the PMST register MP/MC, OVLY, and DROM determine the basic memory configuration as shown in the Table 8.2.

Figure 8.4 indicates the memory map for the program and for the data.

Figure 8.5 indicates the input and output (I/O) space memory map. There are eight Complex Programmable Logic Device (CPLD) registers that are used for software control of various board features. These registers are mapped into the DSP's lower I/O address space starting at address 0x0000. The upper 32 K of the I/O address space is available for daughter-cards.

Address (Hex)	Size (Dec)	Function
0x0000 0x0007	8	CPLD Configuration
0x0008 0x7FFF	32760	Reserved
0x8000 0xFF7F	32640	Daughter Card Access

Fig. 8.5 Input and output memory space map.

The extended memory mapping has been left out. It can be found in [22]. An understanding of the memory mapping is important as it is required during program development in code composer studio.

8.8   The TMS 320CV5416 Development Kit

The TMS320CV5416 DSK is a development system whose block diagram is shown in Figure 8.6. It enables designers to develop and test applications for the TI C544xx DSP. The key features of the system and their functions are summarized below:

(i) TMS320CV5416: A processor that is optimized for low power operation and can operate at clock cycle of up to 160 MHz. It can perform most operations in one clock cycle. It works with a number of on-chip resources to improve functionality and reduce development complexity. It has 128 K words memory, on chip-PLL, Timer, 6 channel Direct Memory Access (DMA) controller, 3 multi-channel buffered serial ports and a 16-bit bus for external memory interface, Figure 8.3.

(ii) SRAM: The DSK has one bank of $64\,K\times16$-bit static RAM that can be expanded to $256\,K\times16$ bit and can run at 160 MHz.

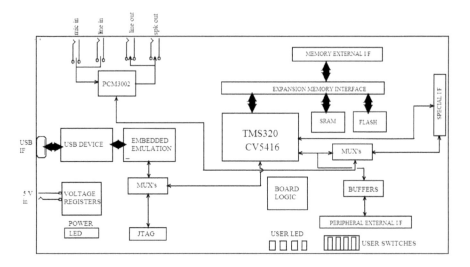

Fig. 8.6 The TMS 320CV5416 development kit [23].

(iii) FLASH: THE DSK has 256×16-bit external Flash memory. This memory space can be used for both data and program.

(iv) EXPANSION CONNECTORS: There are three expansion connectors used for a plug-in daughter-card. The daughter-card adds flexibility to users allowing them to build on the platform and extend its capabilities. The expansion connectors are for the memory, peripherals, and the Host-port interface.

(v) JTAG: There is an embedded JTAG emulator which can be accessed through a universal serial port of a personal computer and also for an external emulator. The JTAG emulator is compatible with the Code Composer debugger. The JTAG emulator can also be used with an external emulator.

(vi) BOARD LOGIC: The DSK uses the programmable logic devices to implement memory mapping, address decoding and assorted glue logic that ties board components together.

(vii) PCM3002 CODEC: The DSK uses a stereo codec for input and output of audio signals. The codec converts the analog input to digital data that can be used by the processor.

The out of the DSP is also converted back to analog by the codec.

(viii) Power supply: The board has voltage regulators that supply 1.6 V DC (core voltage, 3.3 V DC digital and 3.3 V DC analog voltages.

8.9 Code Composer Studio [24]

CCStudio is a window based integrated development environment developed by Texas Instruments. It is a software package that contains basic code generation, debugging and real-time analysis tools. The intention is to make development of DSP applications faster and more efficient. The software development flow would normally follow the following sequence:

(i) Conceptualization and design of the project from project specifications.

(ii) Creating project with source code; normally assembly, C or C++ codes are used.

(iii) Debugging; correcting syntax error and program logic using various debugging features such as probe points, break points, step-by-step execution of commands.

(iv) Analysis: various statistics will be used to analyze utilization of memory, speed of implementation.

It is possible to configure the CCStudio software to work with different hardware and simulator targets. Normally there is a default configuration for a family of targets such as C55xx simulator for the C5000 series and C64xx simulator for the C6000 series. It is also possible to use standard configuration files or modify these to create customized configurations. The procedure for doing configuration can be read in the help files in CCStudio. In order to get access to CCStudio see [25]. The rest of the chapter would not be useful if you have not purchased a TI DSK kit or subscribed for CCStudio.

In order to familiarize oneself with CCStudio the first step would be to go through the CCStudio tutorial in the Help files. This will help you to discover how to create, edit, debug, and run a simple program.

8.9.1   Building and Running a Project

To create a new project one must follow the procedure listed in the Help files under the topic "Creating a New Project." Initially you are guided to open a new project with a name of your choice such as "newproject" in a specified location, e.g., "Myprojects," and selecting a specified target board. When you select "Finish" a project file is created with the name "newproject.pjt" and this is displayed on the "Project View" window. The "newproject.pjt" file stores the project settings and refers to all files that are used in the project.

Following the conceptualization and design of the project the source code is developed and is written in c, c++ or assembly and may be typed in text format in "notepad" or any other text file. If you now open the "File" menu in CCStudio and select the "Creating New Source File" a new page will open where you can paste the notepad text. This is saved as the *.c or *.asm in the "newproject" directory.

The file you have developed must work with a number of files that have different objectives. These are the files that are added in the procedure and they include

(i) Source files (*.c, c++): Normally in c or c++ code and they consist of the new project.

(ii) Configuration files (*.cdb): these are created as New Configuration files from the "File" menu and selecting DSP/BIOS configuration. Here a default simulator, or a standard simulator or even a customized simulator is selected or created. When this source file is added to the project two new files are generated and displayed in the Project View window.

(iii) Linker command files (*.cmd): They consist of

- input files which specify object files, library, or other command files,

- files that specify linker options,

- files that have memory directives that define the target memory configuration and section directives that define how sections are built and allocated.

(iv) Assembly source file (a* or .s*). May be assembled and linked.

(v) Object Library files (.o* or .lib): May only be linked.

(vi) Include file(.h): These files are automatically added to the project list.

When all the files have been added then one has to check on the build options before "Building All." The command "Building All" compiles, assembles, and links all the necessary files in the project. A *.out (in our example a newproject.out) executable file will be created. This file is loaded into the target memory using the command "Load Program." In order to execute the program the command "RUN" from the DEBUG menu is invoked.

8.9.2 Debugging a Program

Debugging of a program involves the use of certain tools provided by CCStudio to detect departure from expected operation and performance of a portion or the complete program. The tools available are listed below:

(i) Probe Points: Using probe points it is possible to inject data from a specified input file into a selected point of the c code or extract data from a selected point of the c code into a specified output file. Such data can also be displayed graphically and analyzed. The procedure is found in the Debug Tools in the Help Files. The operation of Probe points halts the target execution momentarily.

(ii) Break Points: The use of break-points is the same as with probe-points except that with break points the execution is completely halted until when it restarted manually.

(iii) Symbol Browser: The procedure for using symbol browser is found in the HELP file in the Code Composer Tutorial/Debug Tools/Symbol Browser. Simply after Building All and Loading Progam the *.out file is opened. From the TOOLS menu select SYMBOL BROWSER and then select the relevant tabs. The tool allows you to examine the project and its components such as all the source files, all functions, all global variables, user defined data-types, and assembler labels and directives.

(iv) Watch Window: The watch window is a debugging tool where one can define and observe a group of variables. It is selected from the VIEW menu and choosing WATCH WINDOW. It is possible to drag and drop variables from a program into the watch window. In the watch window the values of the variable is displayed.

8.9.3   Data Visualization

Data Visualization is a software tool that is used to display selected data in different graphical formats. The purpose of this tool is to aid in debugging or simply have a means of looking at performance of a software code at an intermediate stage. The tool uses probe points and input and output files in the same manner as when probe points were used for debugging. In order to use this tool a relevant project that generates appropriate data has to be in operation. As a demo in the HELP files the modem project has been used. The procedure for setting data for visualization is given in the HELP files in the Data Visualization folder. It is possible to display the following graphical formats (i) time domain variation of amplitude, (ii) the eye-diagram, (iii) constellation diagram, and (iv) the FFT plots.

8.9.4   Profiling and Optimization of a Program

Profiling an application is to provide a summary of properties and activities of selected or all the functions in the program as the program runs over a specified period of time.

To prepare a project for profiling the commands "BUILD" and "LOAD PROGRAM" must be invoked first. The *.out executable file must be opened. From the profile menu one must then choose START NEW SESSION. Here the profile Session name must be selected. When acknowledged a dialogue box appears with the name and one must select "functions." The source code to be profiled must be opened. The functions to be profiled are selected and dragged into the profile dialogue box. From the menu DEBUG select RUN. After a while select HALT. The dialogue box will show the profile of the functions selected. It is also possible to profile all function and a range containing certain commands in the program.

8.9.5 DSP/BIOS

DSP--BIOS is an important tool that simplifies the design of DSP applications. It consists of three components; real-time kernel, real-time analysis tool and peripheral configuration and management tool. DSP/BIOS has configuration tool that enables the graphical selection of either the kernel or the analysis services and to configure the peripheral devices. The configuration allows efficient utilization of the kernel memory. More information on the tools and how to access them is given below.

(i) Real-time kernel: This tool makes it possible to create and configure DSP/BIOS objects used by your program such as scheduling of tasks, synchronization, etc. It is also possible to configure memory, thread priorities, interrupt-handlers, and Chip support library settings. On the FILE menu choose NEW and select DSP/BIOS CONFIGURATION. This provides templates that have been provided for various standard and default targets. One can use the default target or select a specific standard target which can be used in its current format or can even be customized.

(ii) Real-time analysis tool: This tool allows you to view the program activity in real time. The tool can be used with any project that contains DSP/BIOS configuration. Real-time analysis is the real-time capture and display of data used for the early detection and diagnosis of system-level bugs. DSP/BIOS provides several mechanisms that allow you to acquire, transfer, and display data in real-time with minimal intrusion to the program's operation. For instance you can observe a thread activity using an execution graph. To open a DSP/BIOS analysis tool, use the DSP/BIOS menu or the Real Time Analysis toolbar within Code Composer Studio.

(iii) Device configuration and Management Tool: Your C, C++, and assembly language programs can call over 150 DSP/ BIOS API functions. API stands for Applications Program Interface. The API functions area called by the program header files. More information can be found on API header files and how and when to include such files in your program on the Content Part of the Help file on DSP/BIOS module header files.

8.9.6   Real-Time Data Exchange

Real-time data exchange (RTDX) allows for the exchange of data between a target board and a host computer without interferring with the application of either of the two. RTDX forms a two way data-pipe between the target and the host client. In fact the pipe can be viewed as a collection of thinner pipes or channels. Data is tagged to a specific channel and this makes it possible to distinguish the various data. Data is transferred at any time asynchronously.

The transfer of data is achieved as follows. The target application sends data to the host by calling functions in the RTDX Target Library. These functions immediately buffer the data and then return. The RTDX Target Library then transmits the buffered data to the host in a way as to not interfere with the target application. The host records the data into either a memory buffer or an RTDX log file, depending on the specified RTDX host recording mode. The recorded data can be retrieved by any host application that is a client of the RTDX host interface. The RTDX host interface is provided as the COM interface.

Similarly data can be transferred from the host to the target as shown in the Figure 8.7.

A number of lessons are provided in lesson S1L1 in CCStudio. To understand the RTDX functionality is advisable to go through the entire tutorial which has been written in increasing order of complexity.

8.9.7   Visual Linker Recipe

The visual linker provides a graphical means to configure system memory. It gives a memory map showing the occupation of your

Fig. 8.7 Data exchange between the host and the target.

application and library object files on the target memory description. The operations on the visual linker are made visible and simple by a simple drag and drop of components into specific memory locations. It provides a visual feedback in color showing areas of optimization. When memory layout is satisfactory it generates an executable *.out file. Migration from existing text linker to the visual linker is achieved using a wizard. One must select from the "Option" menu "Link Configuration" and change from "Text Linker" to "Visual Linker."

8.10 Problems

8.9.1 The first step to follow to be able to use CCS is to go through the CCS tutorial in the Help files reached from the CCS window. It will teach you how to load and run a program, how to make simple changes and the use of DSP/BIOS as a configuration tool [12].

8.9.2 Code Composer IDE will expose you to more advanced concepts including; development of a simple programme, project management, editing, the use of debugging tools, data visualization, profiling, the use of the GEL language and configuration of the target devices. It is not a good plan to rush through this in one afternoon. Spend at least a week, on a full-time basis, to understand and experiment with these concepts and procedures properly [12].

8.9.3 DSP/BIOS Tutorial module will introduce you to a number of concepts such as creating and profiling functions, debugging program behavior (scheduling of functions and multi-threading), analysis of real time schedules and connecting of input and output devices. This portion may take up to a week, on a full-time basis, to understand and experiment with the concepts and procedures in this section [12].

8.9.4 Visual Linker will introduce you to its use and how to create and open a recipe and how to manage the memory space. Take your time to understand and do the experiment [12].

8.9.5 Work on a simple project.
Develop a variable gain audio amplifier using digital signal processing software and hardware.

8.9.6 Work on a challenging project.
Develop a simple speech communication system. On one board you develop the transmitter with source code, an FEC (convolutional encoder), modulator. On the receiver board develop a demodulator, synchronizer, and a Viterbi decoder. At this stage I have left out the transmitter and receiver filters on purpose.

9

Examples of DSK Implementations

9.1 Introduction

DSK stands for Digital Signal Processing kit and represents the DSP development kit. The TMS 320CV5416 development kit has been discussed in Section 2.7 and the Code composer studio used to develop applications has been discussed in Section 2.8. The purpose of this chapter is to develop the application file that is linked to library files and the include file that define the board and some other useful functions that make the application work in TMS320V5416. We will focus on the development of the program in C and show how the various processing algorithms are written. The key processes involve FIR filter and the IIR filter and these will be developed in full. Other applications that we have developed and tested in our laboratory will be included.

9.2 FIR Filter Implementation

The implementation of FIR digital filtering can be done using either of the following two methods:

(i) Sample-by-sample processing: In this process an input sample is processed before another sample is taken in. The processing must be completed and the output sample sent out within one sample interval. Sample by sample processing is suitable for real-time signal processing.

(ii) Block processing: In this process the input is segmented into a number of blocks. Each block is processed separately creating its output. The outputs from the various blocks are concatenated to produce an overall output. Block by block filtering can be implemented using fast convolution implementation that uses the Fast Fourier Transform.

In the coming sections we will discuss the two processing techniques in the implementation of FIR filtering.

9.2.1   Sample by Sample Filtering Process

To demonstrate this processing method a direct form FIR filtering process will be used with the input represented by $x(n)$ and the output by $y(n)$ related by the equation

$$y(n) = x(n) \otimes h(n) = \sum_{k=0}^{N-1} h(k)x(n-k), \qquad (9.1)$$

where $h(n)$ for $0 \leq n \leq N-1$ represents the filter impulse response samples which are also the coefficients of the filter. In the filtering process the coefficients $h(n)$ are constants and are stored in a fixed coefficients buffer. The input $x(n)$ is stored in the input buffer which is refreshed every sampling period. Figure 9.1 shows the coefficients buffer and the input buffer in the first and second sample intervals. It is noted that each input signal is shifted to the next memory address in the second sample interval. The current input sample $x(n)$ is placed in the memory location defined by the first address and the sample $x(n-L+1)$ is discarded.

In every sampling interval the following three things happen in relation to the input signal in order to refresh the buffer

(i) The data in $x(n-L+1)$ is discarded.
(ii) The data in $x(n-i)$ is shifted to $x(n-i-1)$.
(iii) New data from the ADC is shifted to $x(n)$.

The refreshing process can easily be implemented by the following C program

```
{
    int i                   /* Loop Index */
    for (i = L-1; i>0; i - -)   /*Start with last sample and advance
                                  back to first sample */
```

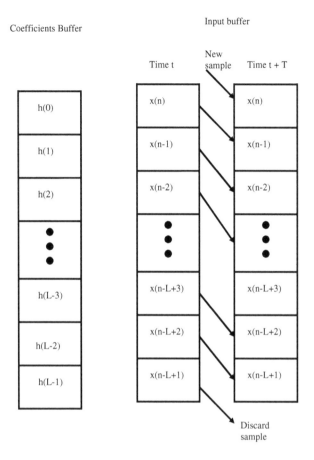

Fig. 9.1 Input and coefficients buffer organization for FIR filtering.

```
{
    xₙ(i) = xₙ(i−1)   /*Shift data to delay it by 1 sample time*/
}
xₙ(0) = input;   /*input new data to obtain the zeroth
                      sample */
return;
}
```

$$x_{n(i)} = x_{n(i-1)} \quad /^*\text{Shift data to delay it by 1 sample time}^*/$$

$$x_n(0) = \text{input}; \quad /^*\text{input new data to obtain the zeroth sample }^*/$$

The process of refreshing the input buffer may be computationally intensive particularly when the number of coefficients is large and the process is not embedded in the hardware. A more efficient method is to load the input signals into a circular buffer. Instead of shifting the input data and holding the addresses constant the opposite is done. The data is kept fixed and the addresses are shifted in the counter-clockwise direction. In fact a pointer used to point to the address of the current input is used. In the next sample only the pointer is shifted to point to the address of the last data which in the current sample time would be discarded. This is the new address of the current input sample. The remaining addresses point to the delayed inputs in the clockwise direction. The positions of the coefficients remain fixed. This is shown in the diagram of Figure 9.2. The advantage of using a circular buffer for the coefficients is the wrap round once all the coefficients have been used.

The refreshing process is therefore implemented as follows:

(i) Initially $x(n)$ is the current input from the ADC and all the previous inputs are zero in line with the initial condition.

(ii) The signal buffer pointer points to address $x(n)$, previous data samples $x(n - i)$ for $1 \leq i \leq L - 1$ are loaded in a clockwise direction.

(iii) The output $y(n)$ is computed using Equation (9.1).

(iv) In the next sample interval at time $t + T$ the pointer is shifted in a counter-clockwise direction. The new sample $x(n)$ is loaded into position of $x(n - L + 1)$. The value of $x(n - L + 1)$ at time t is discarded. The rest of the

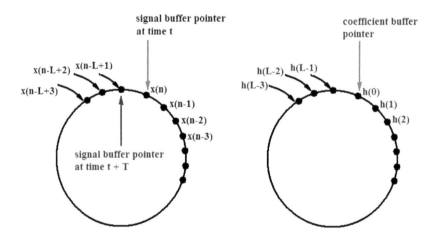

Fig. 9.2 The use of circular buffers in the refreshing processing.

samples remain in a fixed position. Thus the refresh process is achieved by adjusting the pointer position only without physically shifting the data values in the buffer.

The c-code functions that can implement the convolution process and refresh the buffer are given below:

/* This function performs linear convolution between coefficients and input signal samples to produce the sample output y_n */

float fir_filter(float *x, float *h, int filter_order);
/* *x points to the beginning of the array of input samples and *h points to the beginning of the coefficient array*/
```
{
    float yn = 0.0              /* Initialise output yn */
    int i
    for (i = 0; i > N; i ++)    /* Start from 0 to N-1 */
    {
        yn = yn + h(i)*x(i)     /*y(n) = Σ h(l)x(n − l) */
                                    l=0 to N−1
    }
    return yn
```

$$/* y(n) = \sum_{l=0}^{N-1} h(l)x(n-l) \ */$$

/* This function refreshes the input signal samples x(n), x(n-1), x(n-2), x(n-3)..............x(n-N+1) and inputs a new sample */
```
void shift(float *x, int N, float in)   /* N is filter order */
    int i;                              /* Loop index */
    for(i = N; i>0; i--)                /*Start with last sample and
                                        advance back to the first
                                        sample */

    {
        x[i] = x[i − 1];                /* Delay each data sample by
                                        one unit */
    }
    x(0) = input;   /* input new data to obtain the
                        zeroth sample */
    return;
}
```

9.2.2 Block by Block Filtering Process

In this program the input is segmented into blocks of M samples each.

```
void fir_filter_block (float *input, int M, float *h, int N, *output,
float *x)
{
    float yn
    int i, j;
    for ( j = 0; j < M; j++)                /* From first to last sample in
                                               the block */

        {
        x[0] = input[j];                    /*Insert new M data
                                               samples */
        /* FIR Filtering */
    {

        for yn = 0.0, (i=0; i < N; i ++)
        {
            yn+ = h[i]*x[i]                 /* Convolve x(n) with
                                               h(n) */
        }
        output[j] = yn ;
        /* This function refreshes the samples x(n), x(n-1), x(n-2),... ,
        x(n-N +1) */
        void shift(float *x, int N, float input)
        {
            for( i = N - 1; i > 0; i-- )
            {
                x[i] = x[i - 1];            /* Delay each array data
                                               sample by one unit*/

            }
        }
            return;
}
```

The program above written in c have an extension .c and can be used as
source files or functions that are called in applications that require FIR

filtering. The programs can be written in assembly language (with an extension .asm) to make them more efficient as the assembly language is closer to the machine language which is the final format in which the code is used. In its final format before implementation in DSP hardware the program has to be linked with other libraries and include file that will define the board, input and output interface, memory mapping and other functions as described in Section 2.8. The Final program is given in the appendix as Program 9.1.

9.3 IIR Filtering Implementation

The general expression representing IIR filtering is represented by the constant coefficient difference equation given by

$$y(n) = \sum_{i=0}^{L-1} b(i)x(n-i) - \sum_{j=1}^{M-1} a(i)y(n-j) \qquad (9.2)$$

and the transfer function is given by

$$H(z) = \frac{Y(z)}{X(z)} = \frac{\sum_{i=0}^{L-1} b(i)z^{-i}}{1 + \sum_{j=1}^{M-1} a(i)z^{i}}, \qquad (9.3)$$

where $b(i)$ represents the numerator coefficients and $a(i)$ represents the denominator coefficients. In order to calculate $y(n)$ the processor requires the following:

(i) current input $x(n)$,
(ii) L-1 previous inputs,
(iii) M-1 previous outputs,
(iv) L numerator coefficients, and
(v) M-1 denominator coefficients.

The buffer organization will be as shown in Figure 9.3.

Though there are four different memory segments dealing with storage this is much less memory compared to the case of FIR filters. For the same performance the IIR filter has much shorter filter length than

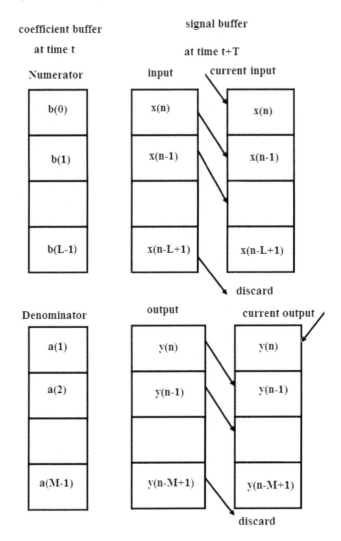

Fig. 9.3 Input, output, and coefficients buffer organization for IIR filtering.

FIR filter. Despite this big advantage of IIR filters circular buffers are still deployed with the intention of making the filtering process faster and more efficient.

The c code below shows how the filtering and the shifting of the signal samples can be achieved. It makes reference to Equation (9.2).

float IIR_filter(float *x, int L, float *a, float *y, int M, float *b);

float y_n
float y_{n1}, y_{n2}
int i,j

y_{n1} = 0.0   /* Initialise y_{n1} */
y_{n2} = 0.0   /* Initialise y_{n2} */
{
for (i = 0; i <= L-1, i++)
    {
        y_{n1} = y_{n1} + x(i) * b(i);   /*Obtain first intermediate filter
                                                   values*/

    }
for (j = 1; j <= M-1, j++)
    {
        y_{n2} = y_{n2} + x(j) * a(j);   /*Obtain second intermediate
                                                   filter values*/

    }
        y_n= y_{n1} − y_{n2};
        return y_n
}

/* This function refreshes the input samples stored in the registers
and reads new input value*/

    void shift1(float *x, int L, float input)
    {
        for(i = L - 1; i > 0; i--)
        {
            x[i] = x[i − 1];   /* Delay each array data sample
                                                   by one unit*/

        }
        x(0) = input
        }

/* This function shifts the output samples stored in the registers*/
    void shift(float *x, int M, float in)

```
{
    for( i = L - 1; i > 0; i-- ).
    {
        y[i] = y[i – 1];            /* Delay each array data sample
                                      by one unit*/

    }
    y(0) = yn                       /* Insert current output into
                                      register*/

}
}
```

The Program 9.1 for the implementation of the FIR filter in the appendix can easily be adapted to implement an IIR filter and will not be reproduced.

9.4   Tone Generation

In this project we will generate a sine-waveform by using a lookup table which has been created using a separate program in MATLAB. We have made all the numbers to be positive. To deal with negative numbers convert decimal to binary and use two's complement representation. The MATLAB program to generate the lookup table is as follows:

```
Theta = 0: 2*pi/50:2*pi;
data = 3168.*(1+ sin(Theta));
data = int16(data);
datah = dec2hex(data);
display(datah);
```

In the project we will generate a sine-wave by reading from a lookup table into an input buffer. We will then scale up the sine-wave by multiplication by a fixed gain and store it in an output buffer. You can display the sine-wave using probe points or send it to the output port. The frequency of the sine-wave is given by $1/(NT)$, where T is the interval between samples and N is the Number of samples in one period. You can reduce the frequency by adding a delay to increase T or increase the number of samples in one period.

```
/*Function: tone.c */
#define N     50                    */ N is the Buffer size*/
const int sinetable[N]=
{0x0C60,  0x0DED,  0x0F74,  0x10EE,  0x1256,  0x13A6,  0x14D9,
0x15E9, 0x16D3, 0x1792, 0x1825, 0x1888, 0x18BA, 0x18BA, 0x1888,
0x1825, 0x1792, 0x16D3, 0x15E9, 0x14D9, 0x13A6, 0x1256, 0x10EE,
0x0F74,  0x0DED,  0x0C60,  0x0AD3,  0x094C,  0x07D2,  0x066A,
0x051A, 0x03E7, 0x02D7, 0x01ED, 0x012E, 0x009B, 0x0038, 0x0006,
0x0006, 0x0038, 0x009B, 0x012E, 0x01ED, 0x02D7, 0x03E7, 0x051A,
0x066A, 0x07D2, 0x094C, 0x0AD3, 0x0C60};

int input_buffer[N];
int output_buffer[N];
int Alpha;                         /* Alpha is the scaling factor */

void main()
{
  int i,j;
  Alpha = 0x50;
  while (1)
  {
    for (i = N-1; i>= 0; i--)
    {
      j = N-1-i;
      output_buffer[j] = 0;        /* Clear the output buffer */
      input_buffer[j] = 0;         /* Clear the input buffer */
    }
    for (i = N-1; i>= 0; i--)
    {
      j = N-1-i;
      input_buffer[j] = sineTable[i];
      out_buffer[j] = Alpha*input_buffer[i];    /* Multiply sine by a
                                                            gain */
      */ Include a statement to send sine to the output port if you
      wish*/
    }
  }
}
```

9.5   Harmonic and Fundamental Component Separator

Project: In this project a square-wave is an input to the TMS320C5416 at any selected frequency. Two filters need to be designed and implemented such that one will isolate the fundamental component of the square wave and feed it into the the right channel output port. The other filter will isolate the third harmonic, amplify it and feed it to the left channel port. It should be possible to display the outputs on an oscilloscope or spectrum analyzer.

Two filters are designed using the fdatool in MATLAB to meet the following specifications:

The Lowpass filter
Passband edge frequency = 1 kHz
Stopband edge frequency = 3 kHz
Maximum passband deviation = 1 dB
Minimum stopband attenuation = 40 dB

Filter Coefficients are placed in h files as follows:

#include "lpfcoeff.h"

const int BL = 8;
const real64_T B[8] = {

0.00979240904553,   -0.05312757622004,   0.01644653849374,
0.5268886286808,   0.5268886286808,   0.01644653849374,
-0.05312757622004, 0.00979240904553
};

The Highpass filter
Passband edge frequency = 3 kHz
Stopband edge frequency = 1 kHz
Maximum passband deviation = 1 dB
Minimum stopband attenuation = 40 dB

#include "hpfcoeff.h"

const int BL = 9;
const real64_T B[9] = {
-0.01127922557944,   0.00850167320085,   0.08571311020913,
-0.2585099772295,   0.3511156226834,   -0.2585099772295,
0.08571311020913, 0.00850167320085, -0.01127922557944
}

The filter magnitude response is shown in the following figure.

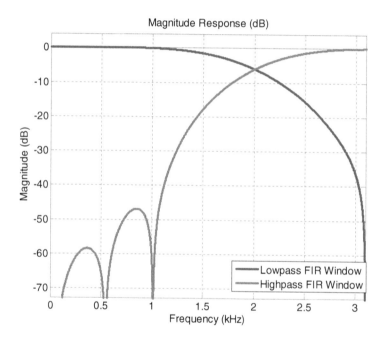

The template for the program harmonics.c that will separate the fundamental at 1 kHz from the 3rd harmonic component is given as

#include "harmonicscfg.h"

#include "dsk5416.h"
#include "dsk5416_pcm3002.h"

```
#include "lpfcoeff.h"        /* will include low pass filter
                                coefficients */
#include "hpfcoeff.h"        /* will include high pass filter
                                coefficients */


/* Configuration setup of registers of PCM3002 Codec */

DSK5416_PCM3002_Config setup = {
0x1FF, / * Set-Up Reg 0 - Left channel DAC attenuation */
0x1FF, /* Set-Up Reg 1 - Right channel DAC attenuation */
0x0, /* Set-Up Reg 2 - Various control such as power-down modes */
0x0  /* Set-Up Reg 3 - Codec data format control */
};

/* To be compatible with pcm3002 read / write, input and output
variables must be declared as Int16 or short int, rather than int. */

Int16 left_input;
Int16 left_output;
Int16 right_input;
Int16 right_output;
Int16 mono_input;

/* UserTask() is scheduled by DSP/BIOS

void UserTask()
{
    DSK5416_PCM3002_CodecHandle hCodec;
    unsigned long i;

    /* Start the codec */

    hCodec = DSK5416_PCM3002_openCodec(0, &setup);
    for ( i = 0 ; i < 20000000 ; i++ )
```

```
{
    /* Read left input channel*/
    while (!DSK5416_PCM3002_read16(hCodec, &left_input));

    /* Output to left output channel */
    while (!DSK5416_PCM3002_write16(hCodec, left_output));

    /* Read right input channel */
    while (!DSK5416_PCM3002_read16(hCodec, &right_input));
    /* Output to right output channel */
    while (!DSK5416_PCM3002_write16(hCodec, right_output));

    /* Where required, generate mono input from left input and
    right input */
    mono_input = stereo_to_mono (left_input, right_input);

  /* Digital Signal Processing goes here */
```

x_{nl} = left_input;

/* write the programme as in Section 9.2.1 which will use the coefficients in lpfcoef.h and obtain the output y_{nl}*/

left_output = y_{nl}

```
    xnr = right_input;
```

/* write the programme as in Section 9.2.1 which will use the coefficients in hpfcoef.h and obtain the output y_{nr}*/
right_output = y_{nr}

```
    }
        /* Finished processing. Close the codec */

        DSK5416_PCM3002_closeCodec(hCodec);


    }
```

```
/* main()                          */

void main()
{
    /* Initialize the board support library      */

    DSK5416_init();
}
```

9.6   The Spectrum Analyzer

In this design we will use the FFT to estimate the power spectral density. A MATLAB function that is used to estimate power spectral density is the periodogram, which is defined for an N-point sequence $x(n)$ as

$$P_N(\omega) = \frac{1}{N}|X(e^{j\omega})|^2 = \frac{1}{N}\left|\sum_{n=0}^{N-1} x(n)e^{-j\omega n}\right|^2.$$

When $x(n)$ is a stationary random process and for large value of N the periodogram oscillates and is not a good estimate of the power spectral density. It can be shown that the mean of the periodogram converges to the power spectral density. For this reason in order to estimate the power spectral density L periodograms of L segments of the sequence $x(n)$ each of length N are obtained and averaged. The estimate of the power spectral density is thus given by

$$S_x(\omega) = \frac{1}{L}\sum_{i=0}^{L-1} P_{N,i}(\omega).$$

One can use the FFT algorithm to compute the periodogram, and hence the power spectral densities, at uniformly spaced frequency points $S_x(\omega_s k/N)$ for $k = 0$ to $N - 1$.

9.6.1   FFT Computation

The program for computing the FFT is available in the board support library as fft.h and fftcpmplx.h and can be accessed as header files.

The task in this project is to design a simple spectrum analyzer that will be able to show the power spectral density of speech using the TMS320C5416 DSK. A guideline on how to write the program is given here. The student should be able to write the various functions that will be used by the main task.

The program guideline for power spectral density estimation is obtained by modifying a program first written by Sakora [ ] to compute the power spectrum of a sequence.

1. Specify files to include
   "psd_estimationcfg.h", "dsk5416.h", "dsk5416_pcm3002.h", "fft.h", "fftcmplx.h".
2. Configure registers of PCM3002 Codec.
3. Define the size N of the FFT (8,16,32,64,128,256, or 512) and define the size of buffers:

input_buffer[N];	Buffer for input samples from codec
COMPLEX y[N];	Variable passed to FFT and modified by FFT output
power_spectrum[N];	Power of each frequency component
total_power[N];	To contain the accumulated FFT after 8 blocks of input
psd[N];	Divide total power by number of input blocks (8)

4. Write Functions to be used by main task

   4.1 Buffers_initialize(): Fill up buffers with zeros to avoid reading of incorrect values in input_buffer[i], power_spectrum[i] , psd[i].

   4.2 shift_and_read(): Shift input buffer by one place. Get latest input into first buffer location after dividing by 4 to limit input range thus avoiding overloading the FFT.

   4.3 Copy_input_to_complex(): Copy from input buffer to complex structure y[N]. When the FFT is performed, the complex structure is overwritten with the return values of the FFT which are complex.

4.4 FFT computation: There is no need to write the programme to compute the FFT as the fft.h and fftcmplx.h are available in board support library.

4.5 Calculate_output_power(): Calculate output power of real and imaginary terms of the FFT. This gives samples of the periodogram.

4.6 Add_and_accumulate(): Add the samples of the periodogram.

4.7 Estimate_power_spectral_density(): simply divides each sample by number of blocks used to obtain the estimate of the power spectral density and stored in array psd[].

```
/*_____*/
/* For compatibility with pcm3002 read / write, the following
variables must be declared as Int16 or short int, rather than int.   */
/*_____*/
Int16 left_input;
Int16 left_output;
Int16 right_input;
Int16 right_output;
Int16 mono_input;


/*_____*/
/* UserTask()                                           */
/*_____*/
/* The main user task.                                  */
/* Note that this task is not called by main(). It is scheduled by
DSP/BIOS                                                 */
/*_____*/

void UserTask()
{
    DSK5416_PCM3002_CodecHandle hCodec;
    long i;
    unsigned int j;
    unsigned int temp;
```

```
/* Start the codec */
hCodec = DSK5416_PCM3002_openCodec(0, &setup);

buffers_initialize();

for ( i = 0 ; i < 120000000 ; i++ )
{
/*Compute FFT for 8 length-N segments of input sequence*/
  for l= 0 ; l< 8 ; l++

/* Read input multiple times and put into receive buffer */
  { for ( j = i+l ; j < i+N ; j++)

    {
    /* Read left input channel */
    while (!DSK5416_PCM3002_read16(hCodec, &left_input));

    /* Output to left output channel */
    while (!DSK5416_PCM3002_write16(hCodec, left_output));

    /* Read right input channel */
    while (!DSK5416_PCM3002_read16(hCodec, &right_input));

    /* Output to right output channel */
    while (!DSK5416_PCM3002_write16(hCodec, right_output));
    /* Copy inputs straight to outputs */
    left_output = left_input;
    right_output = right_input;

    /* Generate mono signal from two input signals */
    mono_input = stereo_to_mono(left_input, right_input);

    /* Read next value into straight buffer */
    shuffle_and_read (mono_input);
      }

    /* Copy input samples from receive buffer to y[] */
    copy_input_to_complex();
```

```
    /* Perforn FFT for specified number of points */
    FFT( y[], N );
    }
```

```
/* Determine power of outputs in y[] */
calculate_output_power( y, &power_spectrum[]); use this
```
statement to compute periodogram

```
    /* Add and accumulate power for 8 segments */

    add_and_accumulate_outpower(power_spectrum, sum_output[]);
    }
```

```
/*Estimate power spectral density

estimate_power_spectral_density(sum_output[], psd[])
```

```
}
```

```
    /* Finished processing. Close the codec */
    DSK5416_PCM3002_closeCodec(hCodec);
```

```
}
/*_____*/
/* main()                                        */
/*_____*/
```

```
void main()
{
    /* Initialize the board support library      */

    DSK5416_init();
    /* All other functions are scheduled by DSP/BIOS */
}
```

The power spectral density can be viewed by observing the psd buffer using a probe.

9.7 The Scrambler

9.7.1 Introduction to the Scrambler

A scrambler is a telecommunication device that can be used to make information coming out of a transmitter unintelligible. In order to be able to interpret information the receiver has to have a descrambler. A scrambler is normally applied to analog systems while an encryptor, a device with the same objectives, is applied to digital systems. A scrambler can also be applied to telecommunication system for a different purpose; to randomize an input data. When a bit stream consists of a long sequence of 1s or 0s the clock in the transmitter fails to track the receiver clock and the system loses synchronization. A scrambler can be deployed to change a long string of 1s or 0s to a more random sequence and hence aid in synchronization.

Scramblers can be classified into two classes; additive scramblers and multiplicative scramblers. Additive scramblers use-modulo two addition to transform the input data stream and in order to achieve synchronization between the transmitter and the receiver a sync-word is used. The sync word has a unique pattern which is placed at the beginning of each frame and is known at the receiver. The receiver searches for it and determines the beginning of incoming frame. Multiplicative scramblers have been named so because in order to determine the scrambled output they perform a multiplication between the input signal and the scrambler transfer function in the z-domain. They do not need a sync-word for frame synchronization and that is why they are referred to as self-synchronizing scramblers. Figure 9.4 shows an ITU-T standard multiplicative scrambler and descrambler defined by the polynomial $1 + x^{18} + x^{23}$.

In general the scrambler output is given by

$$y(n) = \sum_{k=k}^{N} h(k)x(n-k) \tag{9.4}$$

and the descrambler output is given by

$$x(n) = \sum_{i=1}^{N} h(i)y(n-i). \tag{9.5}$$

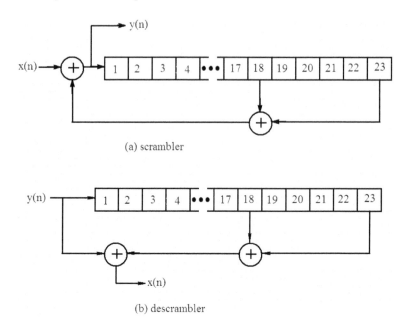

(a) scrambler

(b) descrambler

Fig. 9.4 A standard ITU-T scrambler and descrambler.

$$\text{For the ITU-T standard h}(i) = \begin{cases} 1 & \text{for } i = 1, 5, 18 \\ 0 & \text{otherwise.} \end{cases} \quad \text{and } N = 23.$$

9.7.2   The Scrambler Implementation

The program to implement the scrambler will be written using the following steps:

Write the programs to implement the following functions:

Buffer-initialise to clear the buffer for the input and output to avoid system reading spurious values.

Shift_and_read function to shift data in register to the next location and read the new input and place it in the first register.

Scramble_input function to scramble the input sequence using Equation (9.4).

9.7.3 The Descrambler Implementation

The program to implement the descrambler will be written using the following steps:

Write the programs to implement the following functions:

> **Buffer-initialise** to clear the buffer for the input and output to avoid system reading spurious values
>
> **Shift_and_read function** to shift data in register to the next location and read the new input and place it in the first register.
>
> **Descramble_input function** to descramble the input sequence using Equation (9.5).

9.8 Echo Generator

9.8.1 Single Echo Generator

A single echo generator would produce a single attenuated replica of the input after a pre-determined delay. Figure 9.5 shows the model of a single echo generator. In order to make the delay distinct the delay must be at least a few tens of milliseconds. In the TMS320C5416 the sampling frequency can be fixed and for the audio signal we can select 10 kHz or a sampling period of 0.1 ms. Each unit delay will correspond to 0.1 ms. Thus to get a distinct echo we can use $N = 100$ to give 10 ms delay.

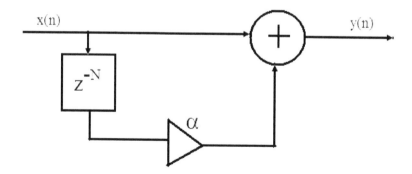

Fig. 9.5 Single echo generator.

The proposed C code will therefore be

```
{
int (i,j)
for i = 0; >500, i++
    {
        y(i) = 0                /*clear output buffer */
        x(i) = 0                /*clear input buffer */
    }
    {
    for j = 500-1; >0; j - -
        {
        x(j) = x(j-1)           /*Shift sample */
        }
    x(0) = input               /* Input new sample */
    yn = x(0) + gain*x(100); /Add echo to input, gain must be less
than 1 */
    output = yn
    }

}
```

9.8.2　Multiple Echo Generator

A system with feedback that will recycle the delayed output over and over and hence produce multiple echoes is shown in Figure 9.6. The output is given by

$$y(n) = x(n - N) + \alpha y(n - N)$$

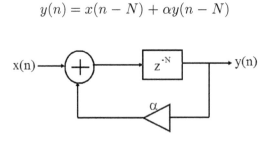

Fig. 9.6 Multiple echo generator.

and the transfer function is given by

$$H(z) = \frac{z^{-N}}{1 - \alpha z^{-N}}.$$

The c code for the generation of multiple echoes is presented below:

Define N 7 /* Define the location of the delay*/

```
{
int (i,j)
for i = 0; >50, i++
    {
        y(i) = 0                    /*clear output buffer */
        x(i) = 0                    /*clear input buffer */
    }
    {
    for j = 50-1; >0; j - -
        {
        x(j) = x(j-1)              /*Shift sample */
        }
    x(0) = input                   /* Input new sample */
    y(i) = x(i-N) + gain*y(i - N); /Add echo to input, gain must be
                                    less than 1 */

    output = y(i)
    }
}
```

9.9 Reverberator

The multiple echo generator does not provide natural reverberations as its magnitude response (which forms a comb filter magnitude response) is not constant at all frequencies. This results in distortion of certain frequency components and make music unpleasant to listen to. To overcome this Schroeder [26, 27] introduced the use of an all-pass structure as shown in Figure 9.7.

The transfer function is given by $H(z) = \frac{z^{-N} - \alpha}{1 - \alpha z^{-N}}$, and the difference equation is given by $y(n) = x(n - N) - \alpha x(n) + \alpha y(n - N)$.

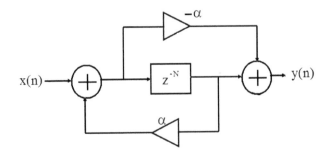

Fig. 9.7 Reverberator using an All-pass structure.

The C-code implementation is given by:

```
# Define N 7 /* Define the location of the delay*/
{
int (i,j)
for i = 0; >50, i++
    {
      y(i) = 0                /*clear output buffer */
      x(i) = 0                /*clear input buffer */
    }
    {
    for j = 50-1; >0; j - -
      {
      x(j) = x(j-1)           /*Shift sample */
      }
    x(0) = input              /* Input new sample */
    y(i) = x(i-N) - gain*x(i) + gain*y(i - N);   /Gain must be less
    than 1 */
    output = y(i)
    }
}
```

To improve on the timber of the output sound one can have a cascade implementation of Figure 9.7.

References

[1] Alan V. Oppenheim and Ronald W. Schafer, *Discrete-Time Signal Processing*, second ed. Prentice Hall Signal Processing Series, 1999. ISBN 0-13-754920-2.

[2] Steven W. Smith, The Scientist and Engineer's Guide to Digital Signal processing, California Technical Publishing, 2007.

[3] E. Oran Brigham, *The Fast Fourier Transform and its Applications*. Englewood Cliffs, NJ: Prentice Hall, 1988, ISBN 0-13-307505-2.

[4] Dragoslav Mitronivić, and Jovan, Kečkić, *The Cauchy Method of Residues: Theory and Applications*, D. Reidel Publishing Company, 1984, ISBN 90-277-1623-4.

[5] Partial Fraction Decomposition Wolfram Mathworld. http://mathworld. wolfram.com/PartialFractionDecomposition.html. Last accessed 10th May 2008.

[6] Butterworth Filters; *Polynomials, Poles and Circuit Elements*, http://www. crbond.com/papers/btf2.pdf. Last accessed 15th May 2008.

[7] S. K. Mitra and J. F. Kaiser, *Handbook for Digital Signal Processing*. John Wiley & Sons, 1993.

[8] John G. Proakis and D. Manolakis, *Digital Signal Processing — Principles, Algorithms and Applications*, Pearson, ISBN 0-13-394289-9.

[9] A Antoniou, *Digital Filters: Analysis, Design and Applications*. New York NY: Mc-Graw-Hill, second ed., 1993.

[10] S. K. Mitra, *Digital Signal Processing: A Computer Based Approach*, McGraw-Hill, second ed., 2001.

[11] IEEE Computer Society, IEEE Standard for Binary Floating-Point Arithmetic, 1985, IEEE Std 754-1985.

[12] DSP Selection Guide, SSDV004S, Texas Instrument Inc., Printed by Southwest Precision Printers, Houston, Texas, 2007.

[13] G. Marven and G. Ewers, *A Simple Approach to Digital Signal Processing*, Texas Instruments, 1994. ISBN 0-904 047-00-8

[14] Michael R. Williams, *A History of Computing Technology*, IEEE Computer Society Press, 1997, ISBN 0-8186-7739-2.

[15] A. W. Burks, H. H. Goldstine, and J. von Neumann, Preliminary discussion of the logical design of an electronic computing instrument, 1963.

[16] J. Eyre and J. Bier, "The evolution of DSP Processors," A BDTI White Paper, Berkely Design Technology Inc., 2000.

[17] ADSP-21xx Processor Architectural Overview, http://www.anlog.com/ processors/adsp/overview/overview.html, Last accessed 10th May 2008.

[18] P. Lapsey, J. River, A. Shoaham, and E. A. Lee, *DSP Processor Fundamentals: Architectures and Features*, Berkely Design Technology Inc., 1996.

[19] BDTi, *Choosing A DSP Processor*. Berkely Design Technology Inc., 1996.

[20] DSP Selection Guide, SSDV004S, Texas Instrument Inc., Printed by Southwest Precision Printers, Houston, Texas, 2007.

[21] TMS320VC 5416, Fixed-Point Digital Signal Processor, Data Manual. Literature Number SPRS0950, Texas Instruments. Revised January 2005.

[22] TMS320C54x DSP. Reference Set. Volume 1: CPU and Peripherals Literature, Number: SPRU131G . Texas Instruments, March 2001.

[23] DSP Starter Kit (DSK) for the TMS320VC5416, Spectrum Digital Incorporated http://www.spectrumdigital.com/product_info.php?&products_id=100. Last accessed, 10th May 2008.

[24] TMS320C5000 Code Composer Studio Manuals, Texas Instruments.

[25] Code Composer studio http://focus.ti.com/docs/toolsw/folders/print/ccstudio. html. Last accessed 10th May 2008. (Access for a trial version or subscription).

[26] MR Schroeder, "Natural sounding artificial reverberation." *Journal of the Audio Engineering Society*, vol. 10, pp. 219–223, 1962.

[27] J. Frenette, "Reducing artificial eeverberation requirements using time variant feedback delay networks, MSc Thesis, University of Miami, Coral Cables, Florida, December 2000.

APPENDIX

Program 2.2

```
% Plotting of the Frequency response of Example 2.2
% Enter the desired length of the DFT
k = input('Enter the number of frequency points = ');
alpha = input('Enter the value of alpha, alpha = ');
% Compute the frequency response
w = -4*pi:pi/k:4*pi;
h = 1./(1 - alpha.*cos(w)+ i*alpha.*sin(w));
subplot(2,1,1);
plot(w/pi, abs(h)); grid
title('Magnitude Spectrum alpha = 0.5');
xlabel('\omega/\pi'); ylabel('Magnitude');
subplot(2,1,2);
plot(w/pi, angle(h)); grid
title('Phase spectrum alpha = 0.5');
xlabel('\omega/\pi'); ylabel('Phase, Radians');
```

Program 2.3

```
% Plotting of the Frequency response of Example 2.3
% Enter the desired length of the DFT
k = input('Enter the number of frequency points = ');
alpha = input('Enter the value of alpha less than 1,
alpha = ');
% Compute the frequency response
w = -4*pi:pi/k:4*pi;
h = 1./(1 - alpha.*cos(w)+ i*alpha.*sin(w));
```

```
subplot(2,1,1);
plot(w/pi, abs(h)); grid
title('Magnitude Spectrum alpha = 2');
xlabel('\omega/\pi'); ylabel('Magnitude');
subplot(2,1,2);
plot(w/pi, angle(h)); grid
title('Phase spectrum alpha = 2');
xlabel('\omega/\pi'); ylabel('Phase, Radians');
```

Program 2.4

```
% Plotting of the Frequency response of Example 2.4
% Enter the desired length of the DFT
k = input('Enter the number of frequency points = ');
alpha = input('Enter the value of alpha greater than1,
alpha = ');
% Compute the frequency response
w = -4*pi:pi/k:4*pi;
h = 1./(1 - alpha.*cos(w)+ i*alpha.*sin(w));
subplot(2,1,1);
plot(w/pi, abs(h)); grid
title('Magnitude Spectrum alpha = 1');
xlabel('\omega/\pi'); ylabel('Magnitude');
subplot(2,1,2);
plot(w/pi, angle(h)); grid
title('Phase spectrum alpha = 1');
xlabel('\omega/\pi'); ylabel('Phase, Radians');
```

Program 6.1

```
% Design a Digital Filter using the Bilinear
Transformation method
% Example 6.1
fp = 10;
% sampling frequency Fs = 200
Fs = 200;
Wp = fp*2*pi/Fs;
```

```
fs = 60;
Ws = fs*2*pi/Fs;
Rp = 0.5;
Rs = 20;
% Pre-warp the normalised digital frequencies to abtain
the analogue
% filter frequencies
Omegap = tan(Wp/2)
Omegas = tan(Ws/2)
[N, Wc] = buttord(Omegap, Omegas, Rp, Rs, 's');
disp('Cut-off frequency of the analogue filter'); disp(Wc);
disp('Order of the analogue filter'); disp(N);
[B, A] = butter(N, Wc,'s');
tf(B,A)
% Apply a bilinear transformation
[Num,Den] = bilinear(B, A, 0.5);
% Display transfer function of digital filter (must enter
sampling time)
tf(Num, Den, 0.005)
omega= [0:0.1:2*Omegas];
Hanalog = freqs(B,A,omega);
subplot(2,1,1)
plot(omega,20*log10(abs(Hanalog)),'m');
title('Butterworth LPF Example 8.1');
xlabel('Frequency in rad/s'); ylabel('Magnitude');
omegad= [0:pi/100:pi];
Hdigital = freqz(Num,Den,omegad);
subplot(2,1,2)
plot(omegad*Fs/(2*pi),abs(Hdigital),'r');
title('Butterworth LPF Example 8.1');
xlabel('Frequency in Hz'); ylabel('Magnitude');
```

Program 6.2

```
% Design a Highpass Chebyshev II Digital Filter using the
Bilinear
```

```
% Transformation method Example 6.2. The sampling
frequency Fs = 120Hz
Fs = 1200;fp = 550;
Wp = fp*2*pi/Fs;
fs = 500;
Ws = fs*2*pi/Fs;
Rp = 1;
Rs = 30;
% Pre-warp the normalised digital frequencies to obtain
the analogue
% filter frequencies
Omega1 = tan(Wp/2);
Omega2 = tan(Ws/2);
% Obtain the specifications of the LPF
Omegap = 1;
Omegas = Omegap*Omega1/Omega2;
[N, Wc] = cheb2ord(Omegap, Omegas, Rp, Rs, 's');
disp('Prototype LPF')
disp('Cut-off frequency Wc = '); disp(Wc);
disp('Order N = '); disp(N);
[B, A] = cheby2(N,Rs,Wc,'s');
tf(B,A)
% Plot magnitude response of the prototype LPF
omega= [0:0.1:2*Omegas];
Hanalog = freqs(B,A,omega);
subplot(2,2,1)
plot(omega,20*log10(abs(Hanalog)),'r');
title('Prototype LPF Example 8.2');
xlabel('Frequency in rad/s'); ylabel('Magnitude');
% Transform the transfer function to an analogue highpass
filter
[BD, AD] = lp2hp(B, A, 2*pi*Wc);
disp('Analogue Highpass filter')
tf(BD,AD)
% Plot magnitude response of the equivalent analogue HPF
omegaa = [0:0.1:2*pi*Wc];
```

```
Hanaloghp = freqs(BD, AD, omegaa);
subplot(2,2,2)
plot(omegaa,20*log10(abs(Hanaloghp)),'b');
title('Analogue HPF Example 8.2');
xlabel('Frequency in rad/s'); ylabel('Magnitude');
% Apply a bilinear transformation
[Num,Den] = bilinear(BD, AD, 0.5);
% Display transfer function of digital filter (enter
sampling time)
disp('Digital Highpass filter')
tf(Num, Den, 0.005)
omegad= [0:pi/100:pi];
Hdigital = freqz(Num,Den,omegad);
subplot(2,2,3)
plot(omegad*Fs/(2*pi),abs(Hdigital),'r');
title('Chebyshev II Example 8.2');
xlabel('Frequency in Hz'); ylabel('Magnitude');
subplot(2,2,4)
zplane(Num, Den);
title('Pole-Zero Plot');
```

Program 6.3

```
% Design a Bandpass Elliptic Digital Filter using the
Bilinear
% Transformation method Example 8.3.
% sampling frequency Fs = 120 KHz
Fs = 120000;
fp1 = 35000; fp2 = 45000;
fs1 = 20000; fs2 = 60000;
Wp1 = fp1*2*pi/Fs; Wp2 = fp2*2*pi/Fs;
Ws1 = fs1*2*pi/Fs; Ws2 = fs2*2*pi/Fs;
Rp = 0.5;
Rs = 40;
% Pre-warp the normalised digital frequencies to obtain
the analogue
```

```
% filter frequencies
Omegap1 = tan(Wp1/2); Omegap2 = tan(Wp2/2);
Omegas1 = tan(Ws1/2); Omegas2 = tan(Ws2/2);
% Check for geometric sysmmetry
if Omegap1*Omegap2>Omegas1*Omegas2
    Omegap1=Omegas1*Omegas2/Omegap2;
    WoSquared =Omegas1*Omegas2;
else Omegas2 = Omegap1*Omegap2/Omegas1;
    WoSquared = Omegap1*Omegap2;
end
Bw = Omegap2 - Omegap1;
% Obtain the specifications of the LPF
Omegap = 1;
Omegas = Omegap*(WoSquared - Omegas1^2)/(Omegas1*Bw);
[N, Wc] = ellipord(Omegap,Omegas,Rp,Rs,'s');
disp('Prototype LPF')
disp('Cut-off frequency Wc = '); disp(Wc);
disp('Order N = '); disp(N);
[B, A] = ellip(N,Rp,Rs,Wc,'s');
tf(B,A)
% Plot magnitude response of the prototype LPF
omega = [0:0.1:2*Omegas];
Hanalog = freqs(B,A,omega);
subplot(2,2,1)
plot(omega,20*log10(abs(Hanalog)),'g');
title('Prototype LPF Example 8.2');
xlabel('Frequency in rad/s'); ylabel('Magnitude');
% Transform the transfer function to an analogue bandpass
filter
[BD, AD] = lp2bp(B, A, Wc, Bw);
disp('Analogue Bandpass filter')
tf(BD,AD)
% Plot magnitude response of the equivalent analogue BPF
omegaa = [0.1:0.1:5];
Hanalogpb = freqs(BD, AD, omegaa);
subplot(2,2,2)
```

```
plot(omegaa,20*log10(abs(Hanalogpb)),'m');
title('Analogue BPF Example 8.3');
xlabel('Frequency in rad/s'); ylabel('Magnitude');
% Apply a bilinear transformation
[Num,Den] = bilinear(BD, AD, 0.5);
% Display transfer function of digital filter (must enter
sampling time)
disp('Digital Bandpass filter')
tf(Num, Den, 0.005)
omegad= [0:pi/100:pi];
Hdigital = freqz(Num,Den,omegad);
subplot(2,2,3)
plot(omegad, abs(Hdigital),'r');
title('Elliptic Bandpass Filter Example 8.3');
xlabel('Frequency in Hz'); ylabel('Magnitude');
subplot(2,2,4)
zplane(Num, Den);
title('Pole-Zero Plot');
```

Program 6.4

```
% Design a Bandstop chebyshev II Digital Filter using the
Bilinear
% Transformation method Example 8.4.
% sampling frequency Fs = 2.8 KHz
Fs = 2800;
fp1 = 700; fp2 = 1200;
fs1 = 800; fs2 = 1000;
Wp1 = fp1*2*pi/Fs; Wp2 = fp2*2*pi/Fs;
Ws1 = fs1*2*pi/Fs; Ws2 = fs2*2*pi/Fs;
Rp = 0.5;
Rs = 40;
% Pre-warp the normalised digital frequencies to obtain
the analogue
% filter frequencies
Omegap1 = tan(Wp1/2); Omegap2 = tan(Wp2/2);
```

```
Omegas1 = tan(Ws1/2); Omegas2 = tan(Ws2/2);
% Check for geometric sysmmetry
if Omegap1*Omegap2>Omegas1*Omegas2
    Omegap1 = Omegas1*Omegas2/Omegap2;
    WoSquared = Omegas1*Omegas2;
else Omegas2 = Omegap1*Omegap2/Omegas1;
    WoSquared = Omegap1*Omegap2;
end
Bw = Omegas2 - Omegas1;
% Obtain the specifications of the LPF
Omegas = 1;
Omegap = Omegas*(Omegap1*Bw)/(WoSquared - Omegap1^2);
[N, Wc] = cheb2ord(Omegap,Omegas,Rp,Rs,'s');
disp('Prototype LPF')
disp('Cut-off frequency Wc = '); disp(Wc);
disp('Order N = '); disp(N);
[B, A] = cheby2(N,Rs,Wc,'s');
tf(B,A)
% Plot magnitude response of the prototype LPF
omega= [0:0.1:2*Omegas];
Hanalog = freqs(B,A,omega);
subplot(2,2,1)
plot(omega,20*log10(abs(Hanalog)),'g');
title('Prototype LPF Example 8.3');
xlabel('Frequency in rad/s'); ylabel('Magnitude');
% Transform the transfer function to an analogue bandpass
filter
[BD, AD] = lp2bs(B, A, Wc, Bw);
disp('Analogue Bandstop filter')
tf(BD,AD)
% Plot magnitude response of the equivalent analogue BPF
omegaa = [0.01:0.1:2];
Hanalogbs = freqs(BD, AD, omegaa);
subplot(2,2,2)
plot(omegaa,20*log10(abs(Hanalogbs)),'m');
title('Analogue BSF Example 8.3');
```

```
xlabel('Frequency in rad/s'); ylabel('Magnitude');
% Apply a bilinear transformation
[Num,Den] = bilinear(BD, AD, 0.5);
% Display transfer function of digital filter (enter
sampling time)
disp('Digital Bandstop filter')
tf(Num, Den, 0.005)
omegad= [0:pi/100:pi];
Hdigital = freqz(Num,Den,omegad);
subplot(2,2,3)
plot(omegad*Fs/(2*pi), abs(Hdigital),'r');
title('Chebyshev II Bandstop Filter Example 8.3');
xlabel('Frequency in Hz'); ylabel('Magnitude');
subplot(2,2,4)
zplane(Num, Den);
title('Pole-Zero Plot');
% Obtain second order section for cascade implementation
Hsos = tf2sos(Hdigital, g)
```

Program 7.1

```
% coefficient Quantisation Effects on the Frequency
response of a
% Direct Form IIR filter
clf;
[N,Wn] = ellipord(1, 2.8618058, 1, 40,'s');
[B,A] = ellip(N,1,40, Wn,'s');
[BT,AT] = lp2bp(B,A, 1.1805647, 0.777771);
[num,den] = bilinear(BT,AT,0.5);
[h,w] = freqz(num,den,512); g = 20*log10(abs(h));
bq = a2dT(num,6); aq = a2dT(den,6);
[hq,w] = freqz(bq,aq,512); gg = 20*log10(abs(hq));
plot(w/pi, g, 'b', w/pi, gg, 'r:'); grid
title('Impact of Coefficient Quantisation');
axis([0 1 -80 5]);
xlabel('\omega/\pi'); ylabel ('Gain, dB');
```

```
pause
zplane(num,den);
title('Case with Unquantised Coefficients');
pause
zplane(bq,aq);
Title('Case with Quantised Coefficients');
function beq = a2dT(d,n)
% BEQ = A2DT(D,N) generates the decimal equivalent beq of
the binary
% representation of the decimal number D with N bits for
the magnitude part
% obtained by truncation.
%
m = 1; d1 = abs(d);
while fix(d1)>0
    d1 = abs(d)/2^m;
    m = m + 1;
end
beq = fix(d1*2^n);
beq = sign(d).*beq.*2^(m-n-1);

function beq = a2dR(d,n)
% BEQ = A2DR(D,N) generates the decimal equivalent beq of
the binary
% representation of the decimal number D with N bits for
the magnitude part
% obtained by rounding.
%
m = 1; d1 = abs(d);
while fix(d1)>0
    d1 = abs(d)/2^m;
    m = m + 1;
end
beq = fix(d1*2^n + 0.5);
beq = sign(d).*beq.*2^(m-n-1);
```

Program 7.2

```
% program for scaling a discrete-time system
% Input filter coefficients.
disp('Coefficients of first cascade section');
disp('aii are numerator and bii are denominator
coefficients');
b00 = input('b00 = '); b01 = input('b01 = ');
b02 = input('b02 = '); a00 = input('a00 = ');
a01 = input('a01 = '); a01 = -a01;
a02 = input('a02 = '); a02 = -a02;
disp('Coefficients of second cascade section');
b10 = input('b10 = '); b11 = input('b11 = ');
b12 = input('b10 = '); a10 = input('a10 = ');
a11 = input('a11 = '); a11 = -a11;
a12 = input('a12 = '); a12 = -a12;
format long
% Computing Scaling factor for signal to first adder of
first cascade
% section
k1 = 1; k2 = 1; k3 = 1;
x1 = 1/k1;
si1 = [0,0]; si2 = [0,0];
varnew = 0; k=1;
while k > 0.000001
    y1 = a01.*si1(1) + a02.*si1(2) + x1;
    x2 = (b00.*y1 + b01.*si1(1) + b02.*si1(2))./k2;
    si1(2) = si1(1);
    si1(1) = y1;
    y2 = x2 + a11.*si2(1) + a12.*si2(2);
    y3 = (y2.*b10 + si2(1).*b11 + si2(2).*b12)./k3;
    si2(2) = si2(1);
    si2(1) = y2;
    varold = varnew;
    varnew = varnew + abs(y1).*abs(y1);
    % Compute approximate L2 norm square
```

```
    k = varnew - varold;
    x1 = 0;
end
k1 = sqrt(varnew);
% Computing Scaling factor for signal to second adder of
first
% cascade section
x1 = 1/k1;
si1 = [0,0]; si2 = [0,0];
varnew = 0; k=1;
while k > 0.000001
    y1 = a01.*si1(1)+ a02.*si1(2)+ x1;
    x2 = (b00.*y1 + b01.*si1(1) + b02.*si1(2))./k2;
    si1(2) = si1(1);
    si1(1) = y1;
    y2 = x2 + a11.*si2(1) + a12.*si2(2);
    y3 = (y2.*b10 + si2(1).*b11 + si2(2).*b12)./k3;
    si2(2) = si2(1);
    si2(1) = y2;
    varold = varnew;
    varnew = varnew + abs(y2).*abs(y2);
    % Compute approximate L2 norm square
    k = varnew - varold;
    x1 = 0;
end
k2 = sqrt(varnew);
% Computing Scaling factor for signal to second adder of
second cascade
section
x1 = 1/k1;
si1 = [0,0]; si2 = [0,0];
varnew = 0; k=1;
while k > 0.000001
    y1 = a01.*si1(1) + a02.*si1(2) + x1;
    x2 = (b00.*y1 + b01.*si1(1) + b02.*si1(2))./k2;
    si1(2) = si1(1);
```

```
    si1(1) = y1;
    y2 = x2 + a11.*si2(1) + a12.*si2(2);
    y3 = (y2.*b10 + si2(1).*b11 + si2(2).*b12)./k3;
    si2(2) = si2(1);
    si2(1) = y2;
    varold = varnew;
    varnew = varnew + abs(y3).*abs(y3);
    % Compute approximate L2 norm square
    k = varnew - varold;
    x1 = 0;
end
k3 = sqrt(varnew);
disp('Scaling factor k1 = '); disp(k1);
disp('Scaling factor k2 = '); disp(k2);
disp('Scaling factor k3 = '); disp(k3);
```

Program 9.1

C Program to Implement FIR filtering.

```
/**********************************************************/
#include <stdio.h> /* Required for functions printf() and
    puts()                   */
/**********************************************************/
/*
/* The next line takes the name of the module, here FIR.c
    and creates FIRcfg.h    */
/**********************************************************/
#include ''FIRcfg.h''
#include ''dsk5416.h''
#include ''dsk5416_pcm3002.h''
#include ''stereo.h''
#include ''FIR_low_pass_filter.h''

/**********************************************************/
/* Configuration setup of registers of PCM3002 Codec    */
/**********************************************************/
```

```
DSK5416_PCM3002_Config setup = {

 0x1FF,  // Set-Up Reg 0 - Left channel DAC attenuation
 0x1FF,  // Set-Up Reg 1 - Right channel DAC attenuation
 0x0,  // Set-Up Reg 2 - Various ctl e.g. power-down modes
 0x0,  // Set-Up Reg 3 - Codec data format control
};

/**********************************************************/
/* For compatibility with pcm3002 read/write, these
   variables must         */
/* be declared as Int16 or short int, rather than int.  */
/**********************************************************/

Int16 left_input;
Int16 left_output;
Int16 right_input;
Int16 right_output;

/**********************************************************/
/* UserTask()                                           */
/**********************************************************/
/* The main user task.                                  */
/* Note that this task is not called by main(). It is
   scheduled by DSP/BIOS                               */
/* the purpose of this section is to read input from the
   two input channels */
/* and to mix them to obtain a mono input              */
/**********************************************************/

void UserTask()
{
  DSK5416_PCM3002_CodecHandle hCodec;
  long l;
  unsigned int switch_value;
  signed int mono_input;
```

```
signed long output;

/*Initialise the input signal registers*/
 define filter_order 51
 int k
 {
     for(k=0;k<filter_order; k++)
     float x[k] = 0.0;
 }
/* Start the codec */
hCodec = DSK5416_PCM3002_openCodec(0, &setup);

/* Read left input channel */

 for (l = 0; l < 12000000; l++)
 {
  while (!DSK5416_PCM3002_read16(hCodec, &left_input));
  while (!DSK5416_PCM3002_read16(hCodec, &right_input));

  /* Merge two inputs from CD player into one */
  mono_input = stereo_to_mono( left_input, right_input );

  /* Perform thr filtering process*/
     float fir_filter(float *x, float *h, int
        filter_order);
      /* *x and *h points to the beginning of the array
         of input samples and input coefficients  */
             {
             float yn = 0.0    /* Initialise output yn */
             int i
             for (i = 0; i > filter_order; i ++)
             {
         yn = yn + h(i)*x(i)   /* Convolve input with
                                     coefficients */

      /*send same output to both channels*/
```

```
        left_output = yn;
        right_output = yn;
        while (!DSK5416_PCM3002_write16(hCodec,
            left_output));
        while (!DSK5416_PCM3002_write16(hCodec,
            right_output));

        /* This function refreshes the input signal
            samples and inputs a new sample */
        void shift(float *x, int Filter_Order,
            float mono_input)

        int i;                           /* Loop index  */
        for(i = filter_order-1; i>0; i--)
         {
         x[i] = x[i - 1]; /* Delay each data sample
            by one unit */
         }
        x(0) = mono_input;
        return;
    }
}
/* Finished processing. Close the codec */
DSK5416_PCM3002_closeCodec(hCodec);
puts("TMS320C5416 DSK program has terminated.\n");
}
/********************************************************/
/* main()                                    */
/********************************************************/
void main()
{
  /* Initialize the board support library           */

  DSK5416_init();

  /* All other functions are scheduled by DSP/BIOS */
```

FIR_low_pass_filter.h
This is the include file containing the filter
coefficients
```
***********************************************************/
/* Lowpass FIR filter with 51 constants for 1200 Hz
   cutoff frequency, */
/* based on 48000 Hz sampling frequency.                 */
/***********************************************************/


#ifndef FIR_low_pass_filter_H
#define FIR_low_pass_pass_H

const signed int FIR_low_pass_filter[] = {
    0x0000, /* H0 */
    0x0000, /* H1 */
    0xFFFF, /* H2 */
    0xFFFE, /* H3 */
    0xFFFE, /* H4 */
    0x0000, /* H5 */
    0x0005, /* H6 */
    0x000F, /* H7 */
    0x0021, /* H8 */
    0x003C, /* H9 */
    0x0063, /* H10 */
    0x0098, /* H11 */
    0x00DD, /* H12 */
    0x0134, /* H13 */
    0x019B, /* H14 */
    0x0214, /* H15 */
    0x029A, /* H16 */
    0x032B, /* H17 */
    0x03C1, /* H18 */
    0x0456, /* H19 */
    0x04E5, /* H20 */
    0x0565, /* H21 */
    0x05D1, /* H22 */
```

```
    0x0622, /* H23 */
    0x0655, /* H24 */
    0x0666, /* H25 */
    0x0655, /* H26 */
    0x0622, /* H27 */
    0x05D1, /* H28 */
    0x0565, /* H29 */
    0x04E5, /* H30 */
    0x0456, /* H31 */
    0x03C1, /* H32 */
    0x032B, /* H33 */
    0x029A, /* H34 */
    0x0214, /* H35 */
    0x019B, /* H36 */
    0x0134, /* H37 */
    0x00DD, /* H38 */
    0x0098, /* H39 */
    0x0063, /* H40 */
    0x003C, /* H41 */
    0x0021, /* H42 */
    0x000F, /* H43 */
    0x0005, /* H44 */
    0x0000, /* H45 */
    0xFFFE, /* H46 */
    0xFFFE, /* H47 */
    0xFFFF, /* H48 */
    0x0000, /* H49 */
    0x0000, /* H50 */
};


#endif


/***********************************************************/
/* End of FIR_low_pass_filter.h                         */
/***********************************************************/
```

Index

9 788792 329127